Advances in African Economic, Social and Political Development

Africa is emerging as a rapidly growing region, still facing major challenges, but with a potential for significant progress – a transformation that necessitates vigorous efforts in research and policy thinking. This book series focuses on three intricately related key aspects of modern-day Africa: economic, social and political development. Making use of recent theoretical and empirical advances, the series aims to provide fresh answers to Africa's development challenges. All the socio-political dimensions of today's Africa are incorporated as they unfold and new policy options are presented. The series aims to provide a broad and interactive forum of science at work for policymaking and to bring together African and international researchers and experts. The series welcomes monographs and contributed volumes for an academic and professional audience, as well as tightly edited conference proceedings.Relevant topics include, but are not limited to, economic policy and trade, regional integration, labor market policies, demographic development, social issues, political economy and political systems, and environmental and energy issues.

More information about this series at http://www.springer.com/series/11885

Nicolas Depetris Chauvin • Guido Porto •
Francis Mulangu

Agricultural Supply Chains, Growth and Poverty in Sub-Saharan Africa

Market Structure, Farm Constraints and Grass-root Institutions

Springer

Nicolas Depetris Chauvin
University of Applied Sciences
Western Switzerland
Haute Ecole de Gestion de Genève
Carouge, Geneva
Switzerland

Guido Porto
Universidad Nacional de La Plata
La Plata, Argentina

Francis Mulangu
African Center for Economic
Transformation
Accra, Ghana

ISSN 2198-7262 ISSN 2198-7270 (electronic)
Advances in African Economic, Social and Political Development
ISBN 978-3-662-53856-2 ISBN 978-3-662-53858-6 (eBook)
DOI 10.1007/978-3-662-53858-6

Library of Congress Control Number: 2017930170

Printed on acid-free paper

This Springer imprint is published by Springer Nature
The registered company is Springer-Verlag GmbH Germany
The registered company address is: Heidelberger Platz 3, 14197 Berlin, Germany

Abstract

This book studies the interplay between market structure and domestic complementary factors in the production and consumption decisions of agricultural families in Africa. We are interested in modelling the production allocation of factors of production to various cash and food crops and in how this allocation depends on competition along the supply chain and on the constraints faced by different types of farmers. The model describes the behaviour of farms, exporters and importers in a simple partial equilibrium setting. In particular, we build three different versions of the model to deal with the three basic scenarios that we face in our empirical work. That is, we build a model to explore the case of cash crop production (mostly for exports). We then adapt this model to deal with the case of a country that is a net exporter of a food crop. Finally, we develop a different version of the model for the case of a country that is a net importer of a food crop. We study changes in market structure and in key parameters of the model that capture various household constraints and institutional access. We analyse the changes in real income of households caused by the hypothetical price changes of cash and food crops predicted by the models' simulations in Tanzania, Malawi, Zambia, Ethiopia, Uganda and Madagascar. In the second part of the book, we evaluate the role of non-market institutions in the context of the constraints found in our case studies. We provide a review of the role of grass-roots institutions in agriculture in Africa, and we study social networking inside and outside communities among both members of grass-roots institutions and non-GRI members in a poor rural district of northern Ghana.

Acknowledgements

We want to warmly thank the African Center for Economic Transformation (ACET) in Ghana that hosts the three authors of the book during most part of the project. George Boateng, David Jaume and Juan Fernandez provided outstanding research assistance. Springer team did superb publishing work.

We have received useful comments and suggestions from numerous participants in seminar presentations, workshops and conferences. These comments helped us to significantly improve the book. We remain, however, responsible for any omission and errors.

Financial support from the DFID-ESRC Growth Research Programme is gratefully acknowledged.

Contents

Chapter 1
Introduction

In Africa, rural poverty is a widespread phenomenon. While Africa has nearly 60 % of the world's arable, uncultivated land, the continent is a major net importer of food and currently one in four people living in Sub-Saharan Africa (SSA) suffers from undernourishment. In most SSA countries, nearly half of the employed work in agriculture and in many of these economies the share of agricultural employment is as high as two-thirds of total employment.

The countries that historically managed to pull out of poverty are those that have been successful in diversifying their economies away from agriculture and other natural resource based activities. The earlier development literature has emphasized the role of agriculture as a facilitator of growth, diversification and poverty reduction (see for instance Johnston and Mellor 1961; Schultz 1964; Christiaensen et al. 2011). A productive agricultural sector can provide non-expensive food and raw materials to start a process of industrialization. However, in Africa, the agriculture sector has so far failed to become an engine of growth and economic transformation for most countries in the continent. With few exceptions, the performance of agriculture in Africa remains unsatisfactory and the sector has not experienced anywhere near the results obtained under the Green Revolution in Asia (Binswanger 2010). This failure has led many local stakeholders to advocate a radical change in the growth strategy of Sub-Saharan Africa and to suggest that countries in the region should import food and shift their focus away of the agriculture sector. Other stakeholders emphasize that while agriculture-led rural transformation may not be a priority everywhere in Africa because of regional heterogeneity of rural situations across and within countries (Dercon and Gollin 2014), it does apply in the case of many regions or countries in Sub-Saharan Africa with a large share of agricultural production and employment.

This recent debate has to be clearly assessed, however. Agriculture mostly comprises tradable commodities. In Africa, international market conditions combine with domestic market configurations in shaping agriculture growth and poverty reduction. The levels of productivity in agriculture in most African countries

N. Depetris Chauvin et al., *Agricultural Supply Chains, Growth and Poverty in Sub-Saharan Africa*, Advances in African Economic, Social and Political Development, DOI 10.1007/978-3-662-53858-6_1

are on the order of one third of those enjoyed by small-holders in Asia. Part of the problem lies in the market structures and in the poor institutions, policies, and infrastructure serving the agriculture sector. Often, the commercialization of the agriculture output is produced along a value chain where intermediaries, exporters, and downstream producers interact with farmers. While in Africa the farming sector is composed mostly of atomistic smallholders, the lower-layers of the value chains are usually dominated by a small number of firms. Farmers may suffer from the non-competitive behavior of other agents along the chain, or be constrained from selling output in markets because transport and other services are not available or are too costly.

While most farmers in Sub Saharan Africa produce food crops for home consumption, some are engaged in high-value export agriculture like tobacco, coffee, cocoa, cotton or tea. Cash crops are a major source of export revenue for a large number of South-Saharan African countries and the livelihood basis for millions of rural households growing those crops. Given their potential key role in development and as a vehicle for poverty reduction, it is not surprising that the policy debate has focused on how to promote the production of these crops, how to create the enabling conditions for smallholders to benefit from the opportunities created by commercial agriculture, and what role should governments play in this process. On the other hand, food crops like maize, rice, millet, sorghum, and soybean are essential for the everyday life of most African farmers as they constitute their dietary base. Growing protests against high food prices in different parts of the developing world, including in Burkina Faso, Cameroon, Cote d'Ivoire, Mauritania, Mozambique, and Senegal have elevated food security as one of the top issues in the international agenda (Conceição and Mendoza 2009). The food imports bill of the world's poorest countries, most of them in Sub Sahara Africa, has considerably increased in recent years, threatening to erase much of the gains in poverty reduction that have been achieved in the last decade. Like in the case of cash crops, food products are also commercialized along a supply chain that includes farmers, silo owners, intermediaries and food processors. In this setting, the structure of the domestic supply chains in staple products affects domestic food prices, agricultural income at the farm level, expenditures, and poverty. Our overall objective is to study market and institutional constraints affecting the further development of the traditional agriculture export sector (cash crops) and the import-substitution agriculture sector (food crops), how this affects poverty and inequality reduction, food security issues, and the development of a competitive agribusiness sector in Sub Saharan Africa.

Traditionally, the literature has focused on how external conditions affect poverty (Winters et al. 2004; Goldberg and Pavcnik 2004, 2007; Nicita 2009; Lederman and Porto 2015). In contrast, the focus of our research agenda breaks new ground by exploring domestic factors and the interplay with international markets. To this end, we elaborate on the work done by Porto et al. (2011) to further explore the role played by the structure of domestic competition in agricultural supply chains. Combining theory, household surveys, and in-depth knowledge of the local context, we use simulation analysis to isolate and quantify the effect of changes in the level of competition in domestic markets, both in food crops and in export crops, on household income. In this setting, we also investigate the role played by household constraints and institutions in agriculture that hinder productivity and market access. The emphasis on the quality of institutions in the development process has recently emerged. Acemoglu et al. (2001, 2002) and Acemoglu and Johnson (2005), for instance, establish causality from better large-scale institutions (like legal and political regimes) to development. Dollar and Kraay (2003), in turn, study how (domestic) institutions affect trade and growth. There is yet another literature that explores how, in low-income countries and especially in rural economies where market failures abound, small local institutions can play a fundamental role as bridges towards economic development and poverty alleviation. Examples that are close to our work include Anderson and Baland (2002) and Besley et al. (1993), on ROSCAs; Garg and Collier (2005), on safety nets and employment; Kranton (1996), on cooperatives; Besley (1995), on risk insurance; Banerjee et al. (2002) on tenancy reform.

In this book, we present results for several Sub-Saharan African countries (Ethiopia, Ghana, Tanzania, Uganda, Madagascar, Malawi and Zambia) and various crops, including cotton, tobacco, coffee, maize, millet, sorghum, cassava, and rice. In Chap. 2, we introduce the model of supply chains. In this model, farmers must decide what to consume and what to produce, given prices and various constraints such as endowments, transport costs, production costs and infrastructure access. In the case of exported cash crops, farmers sell products to oligopsonies, who then do the international trading. In the case of exported food crops, there are oligopsonies in charge of exports, but there is also a domestic residual market of net-consumers of exported food crops. Finally, in the case of imported foodstuff, excess demand is met via international trade, and net-consumers must purchase these agricultural goods from oligopolies. In Chaps. 3–8, we present our six case studies. For each country target, we begin with a description of the household survey used in the analysis. We also review the agricultural supply chains in the selected crops. We describe the basic institutional arrangements and we present information to characterize the supply chains. Finally, we present the results of the simulations. These simulations are comparative static results from the model in Chap. 2, where we study changes in market structure and in key parameters of the model that capture various household constraints and institutional access. We also present here some poverty results stemming from the simulations by combining the predictions of the model with the information from the household surveys. In particular, we analyze the changes in real income of different households caused by the hypothetical price changes of cash and food crops predicted by the models' simulations.

The aim of the second part of the book is to evaluate the role of nonmarket institutions in the context of the constraints found in our case studies. The effort to analysis domestic market conditions with gross-root non-market institutions can help fill a gap in the literature after two decades of reforms in Africa's agricultural marketing (Cadot et al. 2009, 2010). Chapter 9 provides a review of the role of grass-root institutions in agriculture in Africa. Chapter 10 studies social networking inside and outside communities among both members of grassroots institutions and non-GRI members in a poor rural district of northern Ghana.

References

Acemoglu D, Johnson S (2005) Unbundling institutions. J Polit Econ 113:949–995

Acemoglu D, Johnson S, Robinson J (2001) Colonial origins of comparative development: an empirical analysis. Am Econ Rev 91:1369–1401

Acemoglu D, Johnson S, Robinson J (2002) Reversal of fortune: geography and institutions in the making of the modern world income distribution. Q J Econ 117(4):1231–1294

Anderson S, Baland M (2002) The economics of Roscas and intra-household resource allocations. Q J Econ 117(3):963–995

Banerjee A, Gertler P, Ghatak M (2002) Empowerment and efficiency: tenancy reform in West Bengal. J Polit Econ 110(2):239–280

Besley T (1995) Nonmarket institutions for credit and risk sharing in low income countries. J Econ Perspect 9(3):115–127

Besley T, Coate S, Loury G (1993) The economics of rotating savings and credit associations. Am Econ Rev 83(4):792–810

Binswanger, H (2010) Structural transformation and agriculture. ACET

Cadot O, Dutoit L, de Melo J (2009) The elimination of Madagascar's Vanilla marketing board, 10 years on. J Afr Econ 18(3):388–430

Cadot O, Dutoit L, Olarreaga M (2010) Barriers to exit from subsistence agriculture. CEPREMAP Working Papers 1014, CEPREMAP

Christiaensen L, Demery L, Kuhl J (2011) The (evolving) role of agriculture in poverty reduction—an empirical perspective. J Dev Econ 96:239–254

Conceição P, Mendoza R (2009) Anatomy of the global food crisis. Third World Q 30 (6):1159–1182

Dercon S, Gollin D (2014) Agriculture in African development: theories and strategies. Annu Rev Resource Econ 6(1):471–492

Dollar D, Kraay A (2003) Institutions, trade and growth. J Monet Econ 50(1):133–162

Garg A, Collier P (2005) On Kin groups and employment in Africa. Centre for the Study of African Economies, Working paper 95-16, Oxford, UK

Goldberg P, Pavcnik N (2004) Trade, inequality, and poverty: what do we know? Evidence from recent trade liberalization episodes in developing countries. Brookings Trade Forum 2004:223–269

Goldberg P, Pavcnik N (2007) Distributional effects of globalization in developing countries. J Econ Lit 45(1):39–82

Johnston B, Mellor J (1961) The role of agriculture in economic development. Am Econ Rev 4:566–593

Kranton R (1996) The formation of cooperative relationships. J Law Econ Organ 12(1):214–233

Lederman D, Porto G (2015) The price is not always right: on the impacts of commodity prices on households (and countries). World Bank Res Obs 31(1):168–197

Nicita A (2009) The price effect of trade liberalization: measuring the impacts on household welfare. J Dev Econ 89(1):19–27

Porto G, Depetris Chauvin N, Olarreaga M (2011) Supply chains in export agriculture, competition, and poverty in Sub-Saharan Africa. CEPR Press, London

Schultz TW (1964) Transforming traditional agriculture. Yale University Press, New Haven, CT

Winters A, McCulloch N, McKay A (2004) Trade liberalization and poverty: the evidence so far. J Econ Lit 42:72–115

Part I
The Welfare Impact of Agricultural Supply Chains, Market Structure and Farm Constraints

Chapter 2
A Model of Agricultural Supply Chains, Market Structure and Farm Constraints

In this chapter, we introduce the model used to study the interplay between market structure and domestic complementary factors in the production and consumption decisions of agricultural families (farms) in Africa. We are interested in modeling the production allocation of factors of production to various cash and food crops and in how this allocation depends on competition along the supply chain and on the constraints faced by different types of farmers. The model describes the behavior of farms, exporters and importers in a simple partial equilibrium setting. In particular, we build three different versions of the model to deal with the three basic scenarios that we face in our empirical work. That is, we build a model to explore the case of cash crop production (mostly for exports) in Sect. 2.1. This version can be used to study crops such as cotton, coffee, tea, tobacco, cacao, vanilla, etc. We adapt this model to deal with the case of a country that is a net exporter of a food crop in Sect. 2.2. Food crop exports can include any relevant crop in a particular country, namely maize, rice, fish, livestock, etc. Finally, we develop a different version of the model for the case of a country that is a net importer of a food crop in Sect. 2.3. The three versions of the model share common elements, such as the structure of the utility function, the constraints in production, and the market structure, but differ in the way the models are solved to account for export and import prices.

2.1 Cash Crop Exports

2.1.1 Farmers

Consider an economy with a continuum of farmers i, with measure L. Each farmer possesses an endowment e_i of factors of production. It is useful to think about this endowment as a summary indicator of possibly various factors such as land, labor, capital. Farmers can transform this endowment one-to-one into three different

N. Depetris Chauvin et al., *Agricultural Supply Chains, Growth and Poverty in Sub-Saharan Africa*, Advances in African Economic, Social and Political Development, DOI 10.1007/978-3-662-53858-6_2

products: a food crop for auto-consumption (h); a food crop to sell in the market (f); a cash, export crop to be traded with other countries (c).

Food crops can be exchanged in the market at price p^f, which is determined endogenously given total supply and demand. The farmer, though, takes this price as given. Export crops are traded internationally but the farmers cannot export or import goods directly. They instead sell to intermediaries who, after some processing, sell abroad at fixed international prices. The cash crop farmgate price is p^c. We also allow for the presence of transport and transaction costs t_i which may capture lack of access or distance to the market. Farmers earn monetary income d_i from these sales.

Farmer's utility is defined as

$$U_i = \vartheta_i h_i^{\alpha} + d_i,$$

where ϑ_i represents the relative preference of farmer i to produce for the market, after controlling for his endowment, market accessibility and fixed cost to produce crops. This parameter reflects family traditions, including specific knowledge transferred over generations. Importantly, we use it to model different attitudes toward risk and food security. For instance, a farmer may value the own production of food to sustain family needs more than another farmer with similar characteristics. Parameter α measures the decreasing marginal utility of own-food consumption. Farmer's monetary income is d_i, which is equal to

$$d_i = (1 - t_i)\left(p^f - m^f\right)^* f_i + (1 - t_i)(p^c - m^c)^* c_i - F_i,$$

where m^f and m^c are the marginal (unit) costs of producing food crops and export crops respectively, and F_i is the fixed cost of producing crops for export. Note that while the marginal costs are common to all farmers, fixed cost may vary. Differences in fixed costs arise because of differences in setup costs due to various farm constraints and market access constraints, such as missing credit markets, missing credit markets, know-how, scale, etc. These factors create a fixed cost of investment in cash-crop and these costs can vary widely across farmers. To simplify, we assume that marginal costs are instead the same for all farmers. This can be rationalized if farmers use (potentially) the same technology. In principle, the model can accommodate heterogeneity in marginal costs as well as in fixed costs. Given the fixed costs, we assume throughout the analysis that $(p^c - m^c) > (p^f - m^f)$ so that it may be eventually profitable to produce c. In other words, per unit sold, a farmer earns more money with the cash crop than with the food crop. Only a fraction of those farmers, however, will earn enough to cover the fixed costs. Note also that, given the linear technology implied by the constant marginal costs, a farmer will not produce tradable food crops and export crops at the same time. If cash export crops are more profitable, the farmer will allocate all his endowment (net of self-sufficiency requirements) to this crop (and vice versa).

The farmer solves the following optimization problem:

$$\max\ u_i(h, d),$$

subject to

$$d_i = (1 - t_i)\left(p^f - m^f\right)^* f_i + (1 - t_i)(p^c - m^c)^* c_i - F_i,$$
$$e_i = h_i + f_i + c_i.$$

Farmers maximize utility with respect to h_i, f_i and c_i. The optimal production of self-sufficient food h when compared to food sales production f is:

$$\bar{h}_{1i} = \left(\frac{\alpha\vartheta_i}{(p^f - m^f)^*(1 - t_i)}\right)^{1/_{1-\alpha}}.$$

Instead, optimal h when compared with cash crop production c is:

$$\bar{h}_{2i} = \left(\frac{\alpha\vartheta_i}{(p^c - m^c)^*(1 - t_i)}\right)^{1/_{1-\alpha}}.$$

Note that $\bar{h}_2 < \bar{h}_1$ by definition since $(p^c - m^c) > (p^f - m^f)$. The existence of a fixed cost for producing c implies that total cash crop profits should be higher than both specialization in h and production of h and f in the optimum.

The cutoff value of the fixed cost F that would make a farmer indifferent between producing $\bar{h}_{1i}$ of h and $(e_i - \bar{h}_{1i})$ of f and $\bar{h}_{2i}$ of h and the rest $(e_i - \bar{h}_{2i})$ of c is

$$\bar{F}_{1i} = \vartheta_i \bar{h}_{2i}^{\alpha} + (1 - t_i)(p^c - m^c)^*(e - \bar{h}_{2i}) - \vartheta_i \bar{h}_{1i}^{\alpha} + (1 - t_i)\left(p^f - m^f\right)^*(e_i - \bar{h}_{1i}).$$

The value of the fixed cost that would make the farmers indifferent between producing only h and $\bar{h}_{2i}$ of h and the rest $(e_i - \bar{h}_{2i})$ of c is

$$\bar{F}_{2i} = \vartheta_i \bar{h}_{2i}^{\alpha} + (1 - t_i)(p^c - m^c)^*(e_i - \bar{h}_{2i}) - \vartheta_i e^{\alpha}.$$

Given these conditions, it is easy to determine conditions that are consistent with different kinds of production decisions/allocations:

1. If $e_i < \bar{h}_{1i}$ and $e < \bar{h}_{2i}$, the farmer produces $h_i = e_i$.
2. If $e_i < \bar{h}_{1i}$, $e_i > \bar{h}_{2i}$, and $F_i > \bar{F}_{2i}$, the farmer will produce $h_i = e_i$.
3. If $e_i < \bar{h}_{1i}$, $e_i > \bar{h}_{2i}$, and $F_i < \bar{F}_{2i}$, the farmer will produce $c_i = e_i - \bar{h}_{2i}$ and $h_i = \bar{h}_{2i}$.
4. If $e_i > \bar{h}_{1i}$ and $F_i < \bar{F}_{1i}$, the farmer will produce $c_i = e_i - \bar{h}_{2i}$ and $h_i = \bar{h}_{2i}$.
5. If $e_i > \bar{h}_{1i}$ and $F_i > \bar{F}_{1i}$, the farmer will produce $f_i = e_i - \bar{h}_{1i}$ and $h_i = \bar{h}_{1i}$.

These allocations imply the existence of essentially three types of farmers. Some farmers produce only for auto-consumption. These are farmer with very low endowments. For example, a large family living in a farm with little land can only produce some food for self-sufficiency purposes. Other farmers produce some auto-consumption for self-sufficiency and some tradable food crops to sell in the market. This may be surplus food to exchange for money or a different marketable crop. For instance, h may capture a variety of own-consumption crops such as cassava, millet, onions, and white maize, while f may capture hybrid maize sold locally. Finally, a third group of farmers produces for auto-consumption and for the export market. This would be the case of a famer that produces, again, cassava, millet, and perhaps some white maize, but also allocates inputs to cotton, coffee, cacao, tobacco, vanilla, or other similar tradable cash crops (non-food). In this later case, the farmer' endowment must be larger than the threshold $(\bar{h}_2)$so as to have enough production to compensate for the fixed costs incurred to access the export market.

We represent the optimal decision of the farmer based on its endowment in Fig. 2.1.

The farmer chooses the allocation with the highest utility, which depends on several factors. To illustrate, we keep all parameters and factors in the background and focus on the impact of endowments. The curve H corresponds to the increase in farmer's i utility if he produces only h, the line P is the utility of producing h and c, and the curve O is the total utility of producing h and f.

Several observations arise from this graph. Firstly, the marginal utility of h is decreasing, while those of c and f are constant. The intuition behind this is that the law of diminishing marginal utility is stronger for a specific product such as self-sufficiency food than for money in general. In the graph, this means that the marginal utility of producing h is equal to that of producing c and f in points A and D respectively, but it is lower for higher endowments. Points A and D correspond to the endowment thresholds algebraically determined earlier: $\bar{h}_{2i}$ and $\bar{h}_{1i}$. Secondly, if the farmer were to decide to produce c, with endowment $\bar{h}_{2i}$ his/her utility would fall by F_i, which is the fixed cost introduced before. However, from that endowment level onwards, his/her utility increases more than by using the endowment to produce f, increasing by $(1 - t_i)(p^c - m^c)$. This will lead eventually to a point in which the farmer will be indifferent between producing c or f, point C in the diagram.

But the farmer has another option as well: to produce food crop to be sold locally, f. Since selling in the local market has no fixed costs, when the marginal utility of producing f is equal to that of producing h, the household starts producing some f. That point corresponds to endowment level $\bar{h}_{1i}$ and point D. From point D up to point E the farmer will produce $e - \bar{h}_{1i}$ units of f. Point E represents the point in which the higher price the farmer receives for exporting the good compensates the fixed costs the farmer must incur to sell in that market, compared to selling in the local market. After point E, the farmer stops producing f and switches to c, producing $e - \bar{h}_{2i}$ of it.

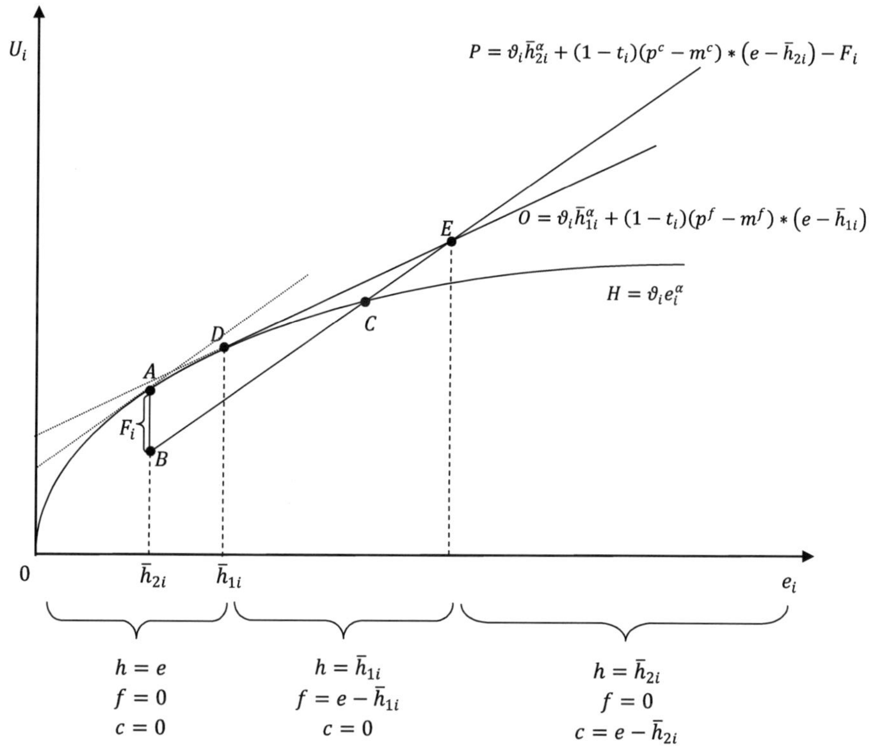

Fig. 2.1 Optimal allocations

To recapitulate, the relevant farmer's i utility is represented in the graph by *ODEP*, and its value, as well as his decision of what to produce, will depend on the endowment e_i at his disposal. As argued above, several factors affect the farmer's decision for a given e_i. Next, we explore graphically the effects of a change in the main parameters of the model.

The parameter ϑ_i accounts for the household's preference to auto-consumption. A larger ϑ_i will increase the marginal utility of producing h for each e, therefore increasing the values of $\bar{h}_{1i}$ and $\bar{h}_{2i}$, as is shown in Fig. 2.2. Farmers that were originally producing f can switch to h if their endowment is between the points D and D', and some farmers with endowment between E and E' will switch from producing some c to produce some f. In the end, farmers originally producing f or c will increase h. In addition, farmers producing c will further switch to f and thus reduce their production of c in $\bar{h}_{2i}(\vartheta'_i) - \bar{h}_{2i}(\vartheta_i)$. In the end, the market supply of c will surely by reduced. The supply of f could either increase or decrease depending on whether or not the farmers switching from form c to f offset the switchers from f to h and the lower f production between $D'$and E.

We now analyze the effects of the change in F in Fig. 2.3. A smaller F_i will reduce the gap AB to AB', affecting the decisions of the farmers with e between E'

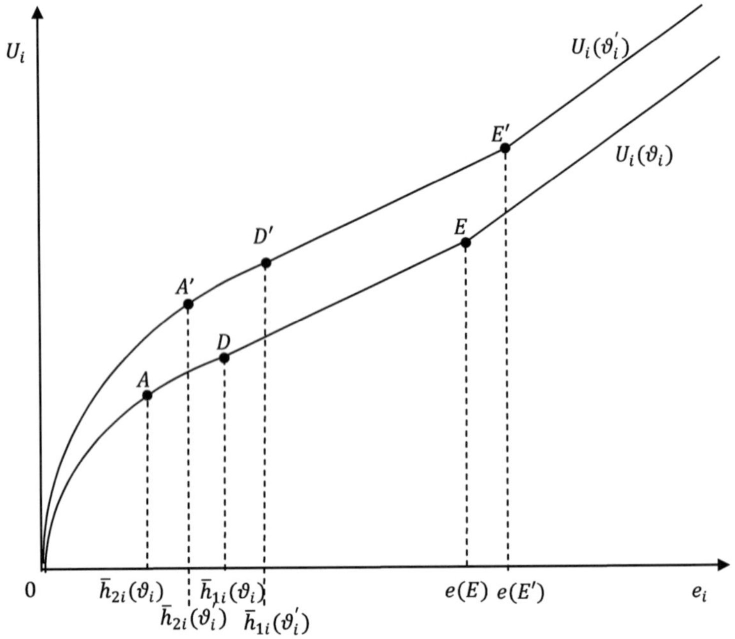

Fig. 2.2 An increase in ϑ_i (from ϑ_i to ϑ'_i)

and E. These farmers will switch from f to h, and they will also reduce h in the amount $\bar{h}_{1i} - \bar{h}_{2i}$. Therefore, lower fixed costs imply a reduction in the total market supply of f and in the production of h, and an increase in the market supply of c. Note that it could be possible to find a F_i small enough so that the farmer will not produce f for any value of e_i. This makes sense: given that $(p^c - m^c) > (p^f - m^f)$, if F_i is low enough, farmers may not produce f at all.

Lastly, we analyze the impacts of changes in the values of t_i and the prices p^c and p^f in Figs. 2.4 and 2.5. These parameters affect the slope of the curves P and O and, consequently, determine the endowment thresholds $\bar{h}_{1i}$ and $\bar{h}_{2i}$, and the points in which the curves H, P and O intercept each other. The effect of an increase in p^c is presented in Fig. 2.4. When the price of c increases from p_0^c to p_1^c it changes the thresholds $\bar{h}_{2i}$ to $\bar{h}_{2i}'$ and $\bar{F}_{2i}$ to $\bar{F}'_{2i}$, which implicitly determine point E, shifting it to E'. The switch leads to more production of f for those farmers that were already producing it ($e > e(E)$) by the amount $\bar{h}_{2i}' - \bar{h}_{2i}$. There will also be switchers, farmers that will adopt the cash exports crops. This is captured by the switch from f to c, when e_i is between $e(E')$ and $e(E)$. These farmers were producing $e_i - \bar{h}_{1i}$ of f and now produce $e_i - \bar{h}_{2i}'$ of c. As expected, thus, an increase in p^c increases the market supply of c.

As shown in Fig. 2.5, an increase in the price of p^f from p_0^f to p_1^f will have opposite effects. Now, point D moves to D' and E to E'. Those farmers between D and D' will switch from h to f, and those between E and E', from c to f. Farmers

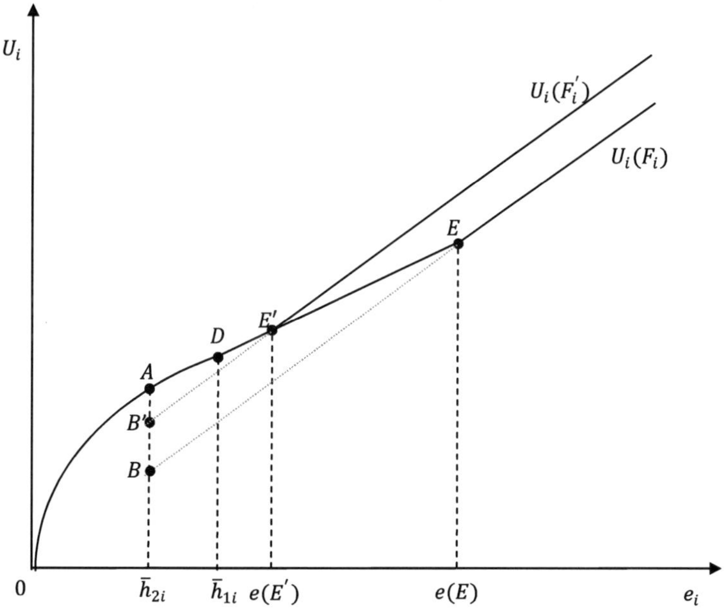

Fig. 2.3 A reduction in fixed costs F_i (from F_i to F'_i)

already producing f will increase their production by $\bar{h}_{1i}\left(p_1^f\right) - \bar{h}'_{1i}(p_0^f)$. The market supply of f is increasing in p^f.

2.1.2 The Farmer Supply of Cash Export Crop

The main purpose of the model is to allow us to derive the supply function of cash export crops. This function will later be combined with a demand for cash export crops to determine equilibrium prices.

To derive the supply, recall that farmers are heterogeneous in potentially many dimensions. We consider four sources of heterogeneity: endowments (e_i), preferences for auto-consumption (ϑ_i), accessibility to markets (t_i),and fixed costs of producing c (F_i). For each of these variables, the heterogeneity is captured by an inherent distribution function. We define $G(e, \vartheta, t, F)$ as the joint distribution of farmers over the different values of e, ϑ, t and F, without any specific functional form assumption (for the moment), with $\int dG(e, \vartheta, t, F) = L$. Using G, we can define $\Omega^c(G, p^c, p^f)$, $\Omega^f(G, p^c, p^f)$ and $\Omega^h(G, p^c, p^f)$ as the farmers that produce crops for export, for the local food market and for auto consumption, respectively.

The supply of cash crop is equal to the sum of the production of all farmers that satisfy conditions 3 or 4 stated above ($e_i < \bar{h}_{1i}, e_i > \bar{h}_{2i}$, and $F_i < \bar{F}_{2i}$; $e_i > \bar{h}_{1i}$ and $F_i < \bar{F}_{1i}$):

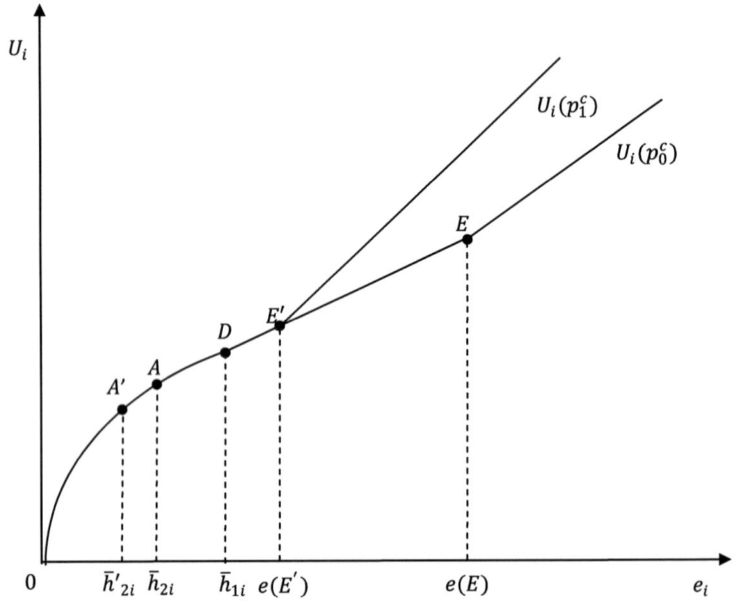

Fig. 2.4 An increase in cash crop price p^c (from p_0^c to p_1^c)

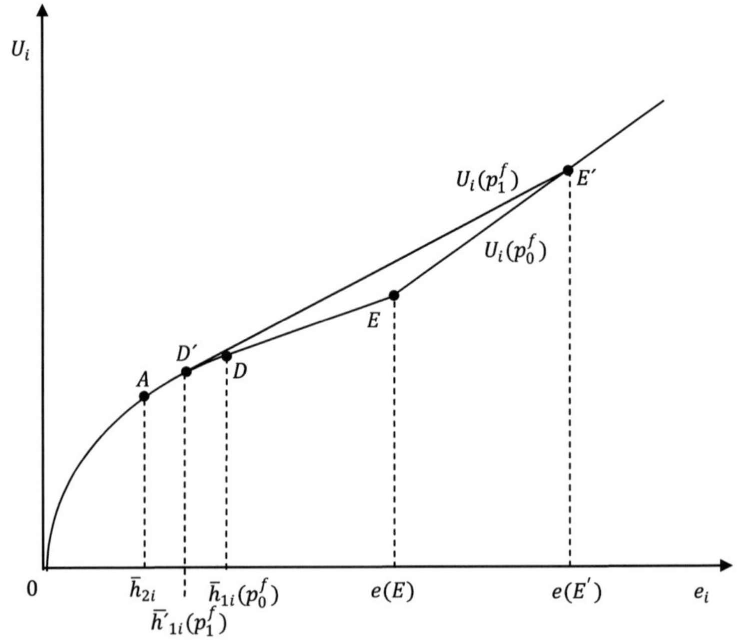

Fig. 2.5 An increase in food sales prices p^f (from p_0^f to p_1^f)

$$S^c(p^c) = \int_{\Omega^c(G,p^c)} (e - \bar{h}_{2i}(p^c))\mathrm{dG}.$$

The supply of food is equal to the sum the farmers' productions who meet condition 5 ($e_i > \bar{h}_{1i}$ and $F_i > \bar{F}_{1i}$):

$$S^f(p^f) = \int_{\Omega^f(G,p^f)} (e - \bar{h}_{1i}(p^f))\mathrm{dG}.$$

Note that

$$\frac{dS^c(p^c)}{d\Omega^c}, \frac{\partial \Omega^c}{\partial p^c}; \frac{\partial \bar{h}_2}{\partial p^c}, \frac{dS^c(p^c)}{dp^c} \geq 0.$$

Similarly,

$$\frac{dS^f(p^f)}{d\Omega^f}, \frac{\partial \Omega^f}{\partial p^f}, \frac{\partial \bar{h}_1}{\partial p^f}, \frac{dS^f(p^f)}{dp^f} \geq 0.$$

The total production of h (denoted by H) is equal to

$$H(p^c, p^f) = \int_{\Omega^h} e\,\mathrm{dG} + \int_{\Omega^f} \bar{h}_{1i}\,\mathrm{dG} + \int_{\Omega^c} \bar{h}_{2i}\,\mathrm{dG}.$$

It is easy to see that

$$S^c(p^c) + S^f(p^f) + H(p^c, p^f) = \int e\,\mathrm{dG}.$$

2.1.3 Exporters

We now turn to the export sector. There are n exporters who sell the crop c at an international price P^c. It is convenient to think about these exporters as firms that do some processing to the farm product. This processing may not necessarily entail complex operations (such as producing high-quality chocolate from cacao). It can be drying coffee beans, cutting tobacco leaves, spinning cotton seeds, or packaging tea leaves or cocoa beans. Exporters buy from farmers at the farmgate market price of p^c. We assume they operate as Cournot oligopsonists. They choose how much quantity to demand from the market at the prevailing price p^c, and they understand and correctly anticipate that their own demand behavior affects p^c.

The problem faced by an exporter is then to maximize profits:

$$\pi\left(P^c, p^c, u_j^c\right) = \max_{c_j} \left(P^c - p^c - u_j^c\right).c_j,$$

where c_j and u_j^c are, respectively, the demanded quantity and the unit cost of production of exporter j of the good c. In principle, exporters may face different marginal costs and this determines the equilibrium market shares.

We look for the equilibrium for the exporters' oligopsony game. Exporters correctly understand and anticipate that the market price p^c depends on their own actions, other exporters' actions, and aggregate supply behavior from farmers. Let $D^c \equiv \sum_{j=1}^n c_j$ denote aggregate demand from exporters, then a given exporter perceives the following problem:

$$\pi\left(c_{k\neq j}, P^c, u_j^c\right) = \max_{c_j} \left(P^c - p^c - u_j^c\right).c_j$$

$$s.t. D^c \equiv c_j + \sum_{k\neq j} c_k$$

The state variables are the international price P^c, and other exporters' actions $c_{k\neq j}$. It can be shown that a sufficient condition for the problem to be concave is that the aggregate supply function $S^c\,(p^c)$ be concave as well, so that $S^{c''}(p^c) < 0$. When the aggregate supply function is concave, the exporters' profit maximization problem will be concave in their choice variable. If the aggregate supply function is not concave, then the problem may not be concave as well. Of course, if the problem is concave then the first order condition $\frac{\partial \pi}{\partial c_j} = 0$ will be necessary and sufficient. Moreover, by the Maximum Theorem under convexity (Stokey and Lucas 1989; Sundaram 1996), the function $c_j(D^{*c})$ is well defined and continuous.

We now turn to the first order conditions. With n exporters, we have

$$c_j^* = \left(P^c - p^c(S^c) - u_j^c\right) \frac{\partial p^c(S^c)}{\partial c},$$

$$D^{*c}(p^c) = \sum_j c_j^*.$$

The equilibrium quantity and price for the export cash crop are determined by the equality of demand and supply, $S^c(p^c) = D^{*c}(p^c)$. The equilibrium is thus characterized by:

$$\int_{\Omega^c(G,p^c)} (e - \bar{h}_{2i}(p^c))dG = D^{*c}(p^c) = \left(nP^c - np^c(S^c) - \sum_{j=1}^n u_j^c\right) \frac{\partial p^c(S^c)}{\partial c}.$$

2.2 Net Food Exports

Here, we adapt the model to study the case of a food crop that is exported by the country. The structure of the model is the same as before. Farmers can produce self-sufficiency food, food crops for sales, and export cash crops. There are intermediaries that buy food from these farmers and sell internationally. The intermediaries compete a-la-Cournot. The model is the same as before. The main difference is that we need to model the local demand for exported food. We begin recapitulating production choices and we then move to demand.

2.2.1 *Production*

In this model, the price of cash crops for exports is assumed to remain constant and we focus our attention on the determination of the price of food for sale (marketable food). Given a price for food sales, the farmer can sell his produce for domestic consumption or for exports. We assume arbitrage and price equalization. To determine equilibrium prices, we need the aggregate net farm supply of food to food exporters. Aggregate gross supply was derived in the previous section and it is given by:

$$S^f(p^f) = \int_{\Omega^f(G, p^f)} \left(e - \bar{h}_{1i}(p^f)\right) \mathrm{dG}.$$

2.2.2 *Domestic Demand*

Since we are working with the case of net food exports, we assume that rural consumers satisfy their own food demand with home food production and thus the rural aggregate supply is just the excess rural production over consumption (net of the resources allocated to the cash export crop). In other words, there is no net food demand in rural areas. This is clearly a simplification but it allows us to succinctly represent the equilibrium in rural food markets.

Aggregate food demand is the sum of urban food demand and of rural food demand. These are slightly different. We begin with urban food demand. We model this as a standard utility maximization problem since we rule out the crop allocation decision. The utility function of the urban consumer i is Cobb–Douglas:

$$U_{ui} = f^{\beta} g^{1-\beta}.$$

Utility is maximized subject to the following budget constraint:

$$fp^f + gp^g = d_i,$$

where g stands for consumption of non-food stuff (goods) with price p^g and d_i is the income of urban households, which is unrelated to agricultural activities (as thus considered exogenous as in the standard utility maximization problem). Individual food demand is $\beta d_i/p^f$. Therefore, the urban demand of food is equal to

$$D_u^f(p^f) = \frac{\beta}{p^f}\int dM(d),$$

where $M(d)$ is the distribution function of income across the urban population.

To model the market food demand of rural consumers, note that the utility for rural households can be written as:

$$U_{ri} = \vartheta_i h_i^\alpha + f^\beta g^{1-\beta}.$$

In this formulation, we assume that food purchases are different from food own-consumption. This could be because these are totally different products (onions and peas in one case, tomatoes and sorghum in another, etc.) or because market foodstuffs comprise different varieties of food. This is clearly a simplification but it allows for a succinct and realistic representation of food markets. The optimum individual consumption of market food f for farmer i is $\beta\frac{d_i}{p^f}$. Recall that money d can take three values: $(1-t_i)(p^f - m^f) * f_i$, if the farmer produces food for sale; $(1-t_i)(p^c - m^c) * c_i - F_i$, if the farmer produces cash crops; and 0, if the farmer only produces auto-consumption. Thus, the aggregate demand for food in rural areas is:

$$\begin{aligned} D_r^f(p^f) = {} & \frac{\beta}{p^f}\frac{\beta}{p^f}(1-t_i)(p^f - m^f)\int_{\Omega^c(G,p^c)}(e - \bar{h}_{2i}(p^c))\mathrm{dG} \\ & + \frac{\beta}{p^f}(1-t_i)(p^f - m^f)\int_{\Omega^f(G,p^f)}(e - \bar{h}_{1i}(p^f))\mathrm{dG}. \end{aligned}$$

Note that farmers producing market staples sell their product in the market at price p^f, and then buy a fraction β at the same price. In our empirical analysis, we will not refer to this process as auto-consumption. This is because our data actually mask heterogeneous goods: the goods the farmers buy are not the same they sell in reality, even if they fit the same category in our taxonomy. Production for the market and consumption from the market with a net exchange of zero is qualitatively very different to auto consumption.

2.2.3 Net Aggregate Supply

At each p^f, there is an urban demand for food, a rural demand for food and an aggregate farm production of food. The gap between demand and supply can be positive or negative, and the difference is absorbed by the external market. If demand is larger than supply, the country is a net importer of a good. Instead, if supply is larger than demand the country is a net exporter of it.

In both cases, net aggregate supply can be defined as

$$NS^f = S^f - D_u^f - D_r^f,$$

so that

$$NS^f(p^f) = \left(1 - \frac{\beta}{p^f}(1 - t_i)(p^f - m^f)\right)\int_{\Omega^f(G,p^f)} (e - \bar{h}_{1i}(p^f))dG$$
$$- \frac{\beta}{p^f}(1 - t_i)(p^c - m^c)\int_{\Omega^c(G,p^c)} (e - \bar{h}_{2i}(p^c))dG - \frac{\beta}{p^f}\int dM(d).$$

It is clear that $\frac{\partial NS^f}{\partial p^f} > 0$, since $\frac{\partial D_u^f}{\partial p^f} < 0, \frac{\partial D_r^f}{\partial p^f} < 0$ and $\frac{\partial S^f}{\partial p^f} > 0$.

In the case of net food exports, we have that $NS^f(p^f) > 0$. The country produces more than it consumes and the excess production is exported. This is done by intermediaries, who buy excess food from farmers and are in charge of the commercialization aboard (and in urban areas). These intermediaries may behave as an oligopoly (as in the case of cash exports). To simplify the reading and the description of the model, we reproduce below the main features of the oligopolistic game.

As before, there are n exporters who sell marketable food f at a fixed international price P^f. They buy from farmers at the internal market price p^f. The oligopoly game is Cournot. Firms (exporters) choose how much quantity to demand from the market at the prevailing price p^f, and they understand and correctly anticipate that their own demand behavior affects p^f.

The problem faced by a food exporter is to maximize profits:

$$\Pi\left(P^f, p^f, u_j^f\right) = \max_{f_j} \left(P^f - p^f - u_j^f\right) f_j$$

where f_j is the quantity of food demanded by exporter j, and u_j^f is the unit cost of production of this exporter (representing, for instance, packaging or processing costs). In principle, exporters may face different marginal costs and this determines the equilibrium market shares. Let $D^{*f} \equiv \sum_{j=1}^{n} f_j$ denote the aggregate food demand from the exporters. A given exporter solves the following problem:

$$\Pi\left(f_{k\neq j}, P^f, u_j^f\right) = \max_{f_j} \left(P^f - p^f - u_j^f\right) f_j$$
$$s.t. D^{*f} \equiv f_j + \sum\nolimits_{k\neq j} f_k$$

The state variables are the international price P^f, and other exporters' actions $f_{k\neq j}$. It can be shown that a sufficient condition for the problem to be concave is that the aggregate net supply function $NS^f(p^f)$ be concave as well, so that $NS^{f''}(p^f) < 0$. If the problem is concave then the first order condition $\frac{\partial \pi}{\partial f_j} = 0$ will be necessary and sufficient. Moreover, by the Maximum Theorem under convexity (Stokey and Lucas 1989; Sundaram 1996), the function $f_j(D^f)$ is well defined and continuous.

We now turn to the first order conditions. With n exporters, we have

$$f_j = \left(P^f - p^f\left(S^f\right) - u_j^f\right) \frac{\partial p^f\left(S^f\right)}{\partial f}$$
$$\Rightarrow D^{*f}\left(p^f\right) = \left(nP^f - np^f\left(S^f\right) - \sum\nolimits_{j=1}^{n} u_j^f\right) \frac{\partial p^f\left(S^f\right)}{\partial f}$$

The equilibrium price is determined by the equality of the exporters demand and the farmers net supply of food, $NS^f(p^f) = D^{*f}(p^f)$.

2.3 Net Food Imports

The model is the same as in Sect. 2.3. The only difference is that in the case of food imports, demand is greater than supply, $D_u^f + D_r^f > S^f$. There is an excess food demand which is satisfied with food imports from abroad.

2.3.1 Production and Domestic Demand

Total food supply is, as before, given by:

$$S^f\left(p^f\right) = \int_{\Omega^f(G, p^f)} \left(e - \bar{h}_{1i}\left(p^f\right)\right) \mathrm{dG}.$$

In turn, urban demand and rural food demands are given by:

$$D_u^f(p^f) = \frac{\beta}{p^f}\int dG(d);$$

$$\begin{aligned} D_r^f(p^f) = {} & \frac{\beta}{p^f}(1-t_i)(p^c-m^c)\int_{\Omega^c(G,p^c)}(e-\bar{h}_{2i}(p^c))dG \\ & +\frac{\beta}{p^f}(1-t_i)(p^f-m^f)\int_{\Omega^f(G,p^f)}(e-\bar{h}_{1i}(p^f))dG. \end{aligned}$$

Net demand is defined as

$$ND^f = D_u^f + D_r^f - S^f$$

$$\begin{aligned} ND^f(p^f) = {} & \frac{\beta}{p^f}\int dM + \frac{\beta}{p^f}(1-t_i)(p^c-m^c)\int_{\Omega^c(G,p^c)}(e-\bar{h}_{2i}(p^c))dG \\ & -\left(1-\frac{\beta}{p^f}(1-t_i)(p^f-m^f)\right)\int_{\Omega^f(G,p^f)}(e-\bar{h}_{1i}(p^f))dG. \end{aligned}$$

It is clear that $\frac{\partial ND^f}{\partial p^f} < 0$, since $\frac{\partial D_u^f}{\partial p^f} < 0, \frac{\partial D_r^f}{\partial p^f} < 0$ and $\frac{\partial S^f}{\partial p^f} > 0$.

As we mentioned above, in this model the country demands more food than it produces. The difference is covered with imports. Imports are brought into the country by intermediaries who buy internationally and sell locally in a potential setting of imperfect competition.

To model this, as before, we assume that there are n importers who buy the food f at an international price P^f. They sell to domestic farmers and urban households at an internal market price p^f. These are Cournot oligopolists. The problem faced by an importer is then to maximize revenues:

$$\Pi\left(p^f, P^f, u_j^f\right) = \max_{f_j}\left(p^f - P^f - u_j^f\right).f_j$$

where f_j is the quantity of food sold by importer j and u_j^f is the unit cost of production (e.g., packaging, distribution, etc.). In principle, importers may face different marginal costs and this determines the equilibrium market shares. Let $S^f \equiv \sum_{j=1}^n f_j$ denote aggregate supply from importers. A given importer solves:

$$\Pi\left(f_{k\neq j}, P^f, u_j^f\right) = \max_{f_j}\left(p^f - P^f - u_j^f\right).f_j$$

$$s.t. S^f \equiv f_j + \sum\nolimits_{k\neq j} f_k$$

With n importers, the first order conditions are:

$$f_j = \left(p^f\left(ND^f\right) - P^f - u_j^f\right)\frac{\partial p^f\left(ND^f\right)}{\partial f}$$

$$\Rightarrow S^f\left(p^f\right) = \left(n\,p^f\left(ND^f\right) - nP^f - \sum_{j=1}^{n} u_j^f\right)\frac{\partial p^f\left(ND^f\right)}{\partial f}$$

In equilibrium,

$$S^f\left(p^f\right) = ND^f\left(p^f\right).$$

2.4 The Solution

The model presented here must be solved numerically. Once a solution is obtained, the equilibrium can be shocked to generate comparative static results that will be used in the welfare analysis in subsequent chapters. In this section, we explain how we calibrate the main parameters of the model and we describe the algorithm used to solve it. As an illustration, we work with the net food export model of Sect. 2.2.

Farmers choose a production allocation and a food demand bundle. Urban households also choose how much to consume of food. There are n Cournot oligopsonist firms that buy food crops from the farmers and sell the surplus in the international market. As we stated before, we need to find the equilibrium where the net domestic supply of food equals the companies' demand: $NS^f(p^f) = D^{*f}(p^f)$.

The first step in the solution of the model is to numerically simulate the allocations of a large number of farmers, based on common and heterogeneous characteristics. The parameters that are common to all farmers are: α_r; α_u; m^f; m^c; p^c ; P^f. The share of food consumed in urban and rural areas is retrieved from the household surveys. Using data from exports and imports, we calculate export and import quantities as well as measure of exports and import prices. These are combined with market information to calculate the ratio of domestic prices to the international price of cash crops. Note that in the case of the net food exporter model of Sect. 2.2 and of the net food imported model of Sect. 2.3, we consider p^c as a fixed parameter that is not affected by changes in the market of f. In this sense, our results capture partial equilibrium effects. As it was also explained, the margin analysis of each crop allows us to compute measures of the price wedges (with respect to international prices) for food crops and thus measures of relative prices.

The heterogeneous parameters that vary across farmers are the endowment (e_i), the transport cost (t_i), the fixed cost F_i and the preference for auto-consumption (ϑ_i). We also need to consider the incomes of urban households (d), used only to obtain the urban demand of f. Endowments in rural areas and income in urban areas are taken from the household surveys. Transport costs are inferred from supplementary information. The preference for auto-consumption is computed from the share of auto-consumption in total household expenditures. Fixed costs are arbitrarily set to the share of producers in the data.

With all these parameters, we can compute $\bar{h}_{1i}$, $\bar{h}_{2i}$, $\bar{F}_{1i}$, and $\bar{F}_{2i}$ for each p^f. These quantities are then used to determine self-sufficiency food consumption $h_i(p^f)$, market food demand $f_i(p^f)$ and cash crop production $c_i(p^f)$. Next, we calculate aggregate food supply $S^f(p^f)$ and the domestic demands $D_u^f(p^f)$ and $D_r^f(p^f)$. Net supply $(NS^f(p^f))$ is equal to $S^f(p^f) - D_u^f(p^f) - D_r^f(p^f)$.

We now need to compute the total food demanded by the oligopsony enterprises j. We have information about the share that each firm has in the market, and we need to compute their marginal cost (u_j^f). For that purpose, we use export and import records to assess the total quantity demanded (D^{*f}) and we use this to solve for the original equilibrium price and the farmer marginal costs using $S^f(p^f) = D_u^f(p^f) + D_r^f(p^f) + D^{*f}$.

Then, we calculate the marginal cost of company j as

$$u_j^f = P^f - p^f(NS^f) - f_j \frac{\partial NS^f(p_0^f)}{\partial p^f}$$

Note that $\frac{\partial NS^f(p^f)}{\partial p^f(ND^f)}$ can be easily calculated since we have already estimated the aggregate net supply $NS^f(p^f)$. We do all this to calibrate the u_j^f compatible with the shares from data and the aggregate demand $D^{*f}(p^f)$.

Given the solution to the model, we can simulate the impacts, especially on prices, of changes in several parameters. This is done by solving the model under the changed parameter configuration to find a price p^f such as $S^f(p^f) = D_u^f(p^f) + D_r^f(p^f) + D^{*f}(p^f)$. As a result, we obtain the equilibrium quantities $h(p^f)$ and $c(p^f)$ produced by the farmers and the $f(p^f)$ consumed by rural and urban households.

The cash export model is Sect. 2.1 is slightly different: we take the value of p^f as fixed and there is no need of estimate the domestic demand for f. We solve for the marginal costs of cash crop production based on the information on price ratios and on the solution of the equality of export supply and demand (given trade flows). Then, we calibrate the marginal cost of the n exporters using

$$u_j^c = P^c - p^c(S^c) - f_j \frac{\partial S^c(p_0^c)}{\partial p^c}$$

With all the calibrated parameters and with the solution to the model, we perform simulations by computing the new equilibrium from $S^c(p^c) = D^{*c}(p^c)$. For the food import demand model of Sect. 2.2, we solve $S^f(p^f) + M = D_u^f(p^f) + D_r^f(p^f)$, or $M = ND^f(p^f)$ and the equation that calibrates the marginal cost of the importers is

$$u_j^f = p^f\left(ND^f\right) - P^f - f_j \frac{\partial ND^f\left(p_0^f\right)}{\partial p^f}$$

Finally, the results from the simulations follow from solving $S^f\left(p^f\right) + S^{*f}\left(p^f\right) = D_u^f\left(p^f\right) + D_r^f\left(p^f\right)$.

References

Stokey N, Lucas RE (1989) Recursive methods in economic dynamics. Harvard University Press, Cambridge, MA

Sundaram RK (1996) A first course in optimization theory. Cambridge University Press, Cambridge

Chapter 3
The Case of Tanzania

This chapter analyzes the case of Tanzania. We describe the household survey (Sect. 3.1), the food and cash crops institutional arrangements (Sect. 3.2) and we present the simulation results from the model (Sect. 3.3) and the household welfare simulations (Sect. 3.4).

3.1 The Household Survey

The household data comes from the 2008 Tanzania National Panel Survey. The dataset contains information on over 16,000 households. Around one third of these households reside in urban areas and two-thirds in rural areas. As it is often the case in Africa, the Tanzanian population is young: 45 % of the sample is <15 years old and over 90 % is under 65 years old. There are slightly more females (52 %) than males (48 %). However, only 25 % of the households are headed by females. On average, household size is 6.77 members per family. In turn, households in rural areas are bigger than in urban areas (7 versus 5.98 members per family respectively).

Figure 3.1 shows the distribution of income. The graph shows the estimated density function of the logarithm of household per capita expenditure at the national level and for urban and rural regions separately. As expected, the density for urban areas lies to the right of the density for rural areas, thus indicating that urban households enjoy, on average, a higher level of expenditure per capita than the rural households. Since the rural sample is bigger, the national distribution of income lies close to the rural density.

We turn now to a description of sources of income and patterns of consumption across households. In Table 3.1, we report consumption patterns for urban and rural regions. We report cash expenditures and the value of auto-consumption, as a share of total expenditures. As expected, the share of auto-consumption is much larger in

N. Depetris Chauvin et al., *Agricultural Supply Chains, Growth and Poverty in Sub-Saharan Africa*, Advances in African Economic, Social and Political Development, DOI 10.1007/978-3-662-53858-6_3

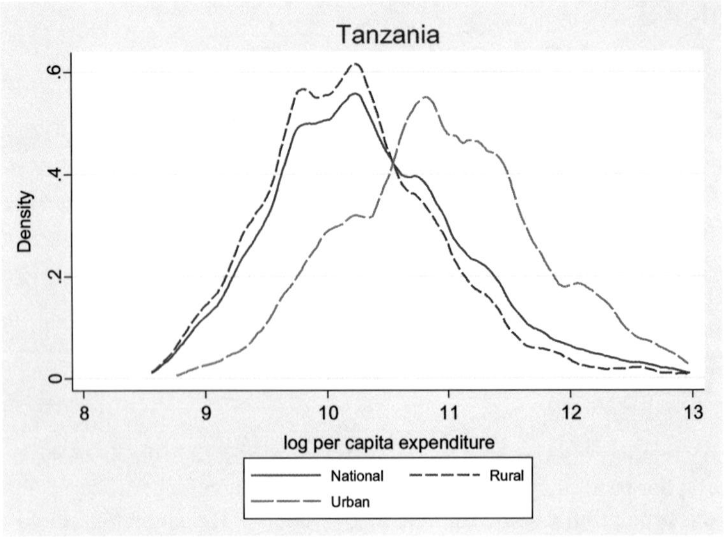

Fig. 3.1 The distribution of income. Density of (log) per capita household expenditure. Source: Tanzania National Panel Survey (2008)

Table 3.1 Budget shares

Tanzania	Total	Rural	Urban
Total consumption per capita	100.0	100.0	100.0
Expenditures	66.9	59.2	93.9
Food	32.3	28.8	44.8
Manufactures	15.2	14.7	16.8
Services	19.4	15.7	32.3
Others	0.0	0.0	0.0
Auto-consumption	33.1	40.8	6.1
Auto-consumption food	33.1	40.8	6.1
Auto-consumption others	0.0	0.0	0.0
Total food consumption	65.4	69.6	50.9
Total crops	39.4	43.2	26.2
Maize	15.7	17.7	8.6
Rice	4.8	4.4	6.4
Livestock	5.9	6.1	5.1
Cassava	3.9	4.8	1.1
Cowpea	4.4	4.9	2.7
Yam	0.3	0.3	0.1
Wheat	1.0	1.0	1.2
Groundnut	1.5	1.8	0.4
Sweet potato	1.9	2.3	0.5

Source: Tanzania national panel survey (2008)

rural areas than in urban areas. In fact, for urban households, 93.9 % of their expenditure is cash spending. For rural households, cash expenditures account for 59.2 % of the total budget, while home-produced expenditures account for the remaining 40.8 %. Since we are interested in food consumption, we can take a close look at aggregate food expenditure, that is, food cash expenditure and food auto-consumption. At the national level, 65.4 % of the Tanzanian budget is allocated to food. This share is larger for rural households (69.6 %) than for urban households (50.9 %). This observation fits with the idea that urban people are richer, who thus spend more on other goods and services than on food. Among food item, the most significant crop in consumption is maize. On average, maize represents 15.7 % of Tanzania's household expenditure (17.7 % of rural expenditure and the 8.6 % of urban expenditure). Rice accounts for 4.8 % of the budget, with slightly higher shares among urban households (6.4 vis-à-vis 4.4 %). Cassava, in turn, accounts for 4.8 % of expenditures in rural areas and for only 1.1 % in urban areas.

In Table 3.2, we show different sources of income. As expected, rural households have low shares of cash income (32.4 %), because their gross income comes mostly from auto-consumption. On the other hand, urban cash income represents 78.4 % of total income. Looking at crop income, maize is the most important crop. It represents 20.7 % of rural household income and 7.4 % of urban household

Table 3.2 Income shares

Tanzania	Total	Rural	Urban
Total income per capita	100.0	100.0	100.0
Incomes	41.1	32.4	78.4
Food (agriculture)	15.1	16.9	7.2
Wage	16.2	8.9	47.7
Enterprises	8.0	5.2	20.2
Transfers	1.8	1.4	3.3
Auto-consumption	58.9	67.6	21.6
Auto-consumption food	58.9	67.6	21.6
Auto-consumption others	0.0	0.0	0.0
Total food income and AC	74.0	84.5	28.8
Total crops	49.2	56.6	17.0
Maize	18.2	20.7	7.4
Rice	4.0	4.5	1.6
Livestock	5.0	5.6	2.4
Cassava	5.4	6.4	1.1
Cowpea	4.4	5.2	1.2
Yam	0.4	0.5	0.0
Wheat	0.8	1.0	0.0
Groundnut	2.6	3.0	1.2
Sweet potato	2.8	3.3	0.2
Cotton	1.4	1.7	0.2
Tobacco	0.7	0.9	0.0
Milk	3.5	3.9	1.6

Source: Tanzania national panel survey (2008)

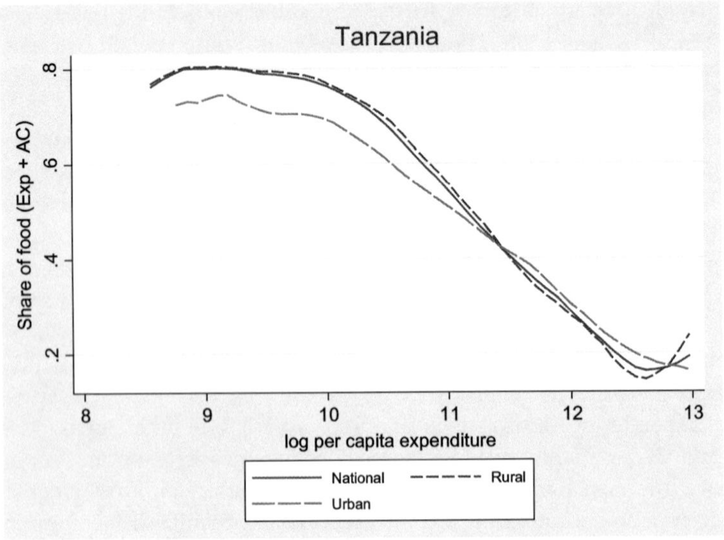

Fig. 3.2 Total food budget share across the income distribution. Source: Tanzania national panel survey (2008)

income. Rice (4.5 %) and cassava (6.4 %) are also relatively important sources of income in rural areas, but not so much in urban areas (1.6 and 1.1 % respectively).

To explore the poverty and welfare impacts of changes in the prices of these commodities, it is important to describe first the patterns of income sources and of expenditure shares across the income distribution. We characterize the distribution of income with the (log) of per capita household expenditure (log pce) and we plot estimates of non-parametric regressions of income and budget shares on log pce.

We begin in Figs. 3.2 and 3.3 with average share of food expenditure (cash expenditure plus auto-consumption) and share of income food (cash agricultural plus auto-consumption). The food share profile slopes steeply downward. In fact, at the bottom of the distribution, around 80 % of the budget is allocated to food, while at the top, only about 20 % is allocated to food. The fact that the curves in Fig. 3.2 slope downward is no more than a manifestation of Engel's law, or its food equivalent that the share of the budget spent on food declines as living standards rise. At the very bottom of the expenditure distribution, rural expenditure is greater than urban expenditure. It noteworthy, however, that as households get richer, these shares converge (and are in fact slightly larger for urban households). Figure 3.3 shows shares of food production on the logarithm of household per capita expenditure. The income share of rural agricultural production is always greater than the urban share. Unlike expenditures, income shares are stable along the log per capita expenditure. From these two figures, we can draw preliminary conclusions about the welfare effects of food price changes. Looking at consumption patterns, price declines will improve welfare conditions relatively more for poor people than for rich people. Looking at Fig. 3.3, lower price will hurt richer (rural) households

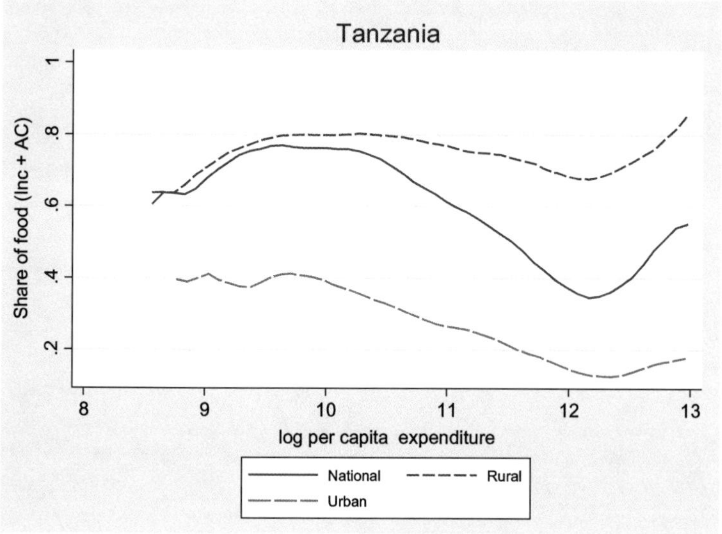

Fig. 3.3 Total food income share across the income distribution. Source: Tanzania national panel survey (2008)

proportionately more than poor households. This illustrates potential differences in the distributional impacts of price changes.

We now take a closer look at the patterns of income and budget shares across the income distribution for the main crops under study in Tanzania (Figs. 3.4, 3.5, 3.6, 3.7, 3.8 and 3.9). Urban budget expenditures in rice are larger than rural shares along the log per capita expenditure (Fig. 3.4). Middle class households spend more on rice in both regions. Looking at income shares in rice (Fig. 3.5), the low and middle classes among urban households show a high income rice share. Among the wealthier households, the rural rice income share is larger than the urban share.

Maize is almost completely consumed by poor people (Fig. 3.6). The rural share of maize is larger at all levels of wellbeing (log per capita expenditure). An increase in the price of maize will affect poor people more than rich people, because their consumption budget in maize is larger. Looking at income shares (Fig. 3.7), we find that poor people in urban regions have larger maize income shares than richer households. This follows because the nonparametric regressions slope down. On the other hand, the negative correlation between the share of maize in income and the level of expenditure is not clear among rural farms.

Fig. 3.8 shows a clear relation between the share expenditures on cassava and the level of expenditure. Cassava consumption tends to decrease when log per capita expenditures rises. For poor people it represents almost 15 % of their budget share, whereas the budget share of rich people is negligible. This is a crop that is almost completely consumed by poor and middle-income households, which are thus more likely to be affected by cassava price changes. Figure 3.9 shows cassava income shares. The relationship with wellbeing is still negative but much less pronounced.

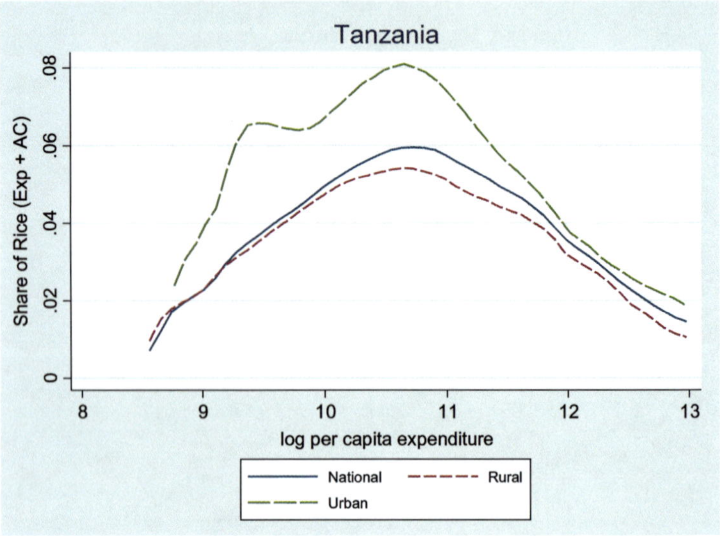

Fig. 3.4 Rice budget share across the income distribution. Source: Tanzania national panel survey (2008)

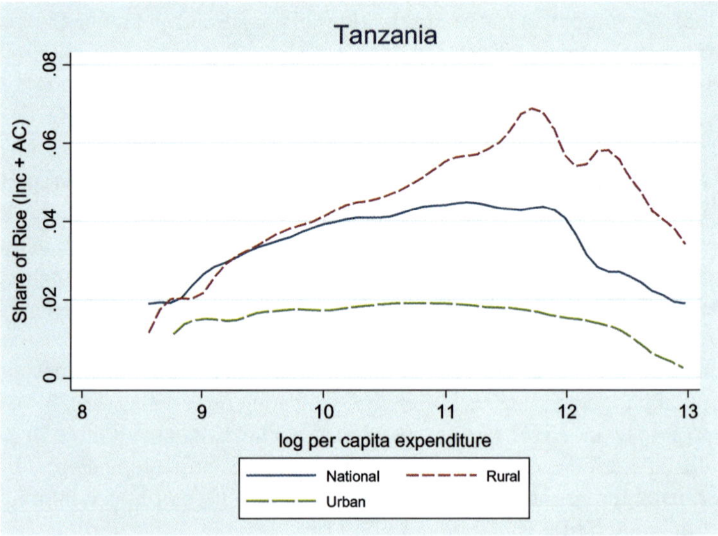

Fig. 3.5 Rice income share across the income distribution. Source: Tanzania national panel survey (2008)

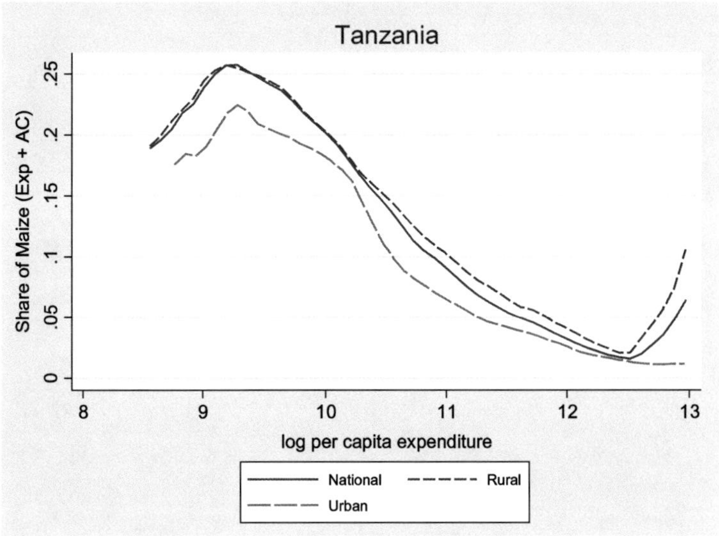

Fig. 3.6 Maize budget share across the income distribution. Source: Tanzania national panel survey (2008)

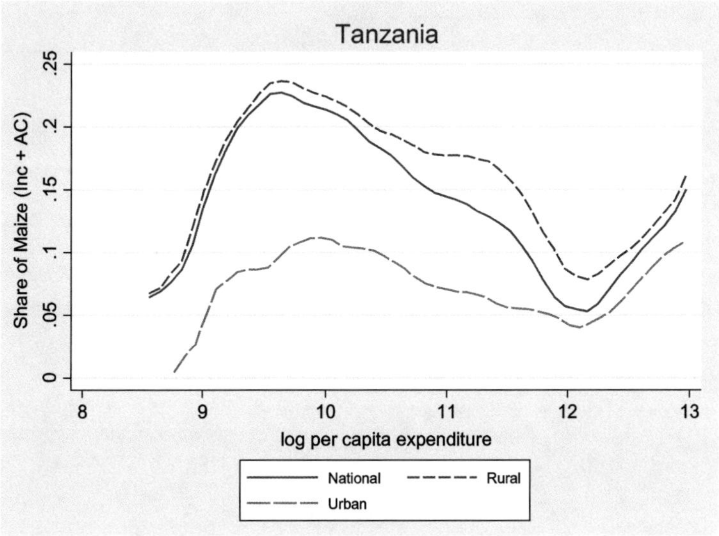

Fig. 3.7 Maize income share across the income distribution. Source: Tanzania national panel survey (2008)

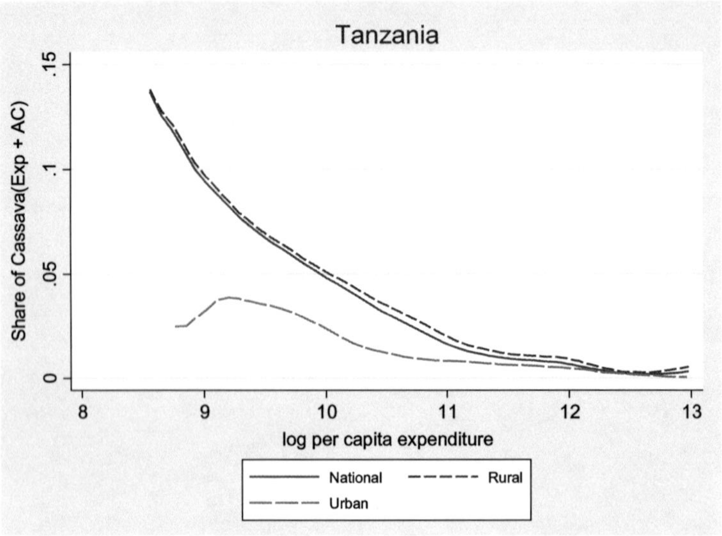

Fig. 3.8 Cassava budget share across the income distribution. Source: Tanzania national panel survey (2008)

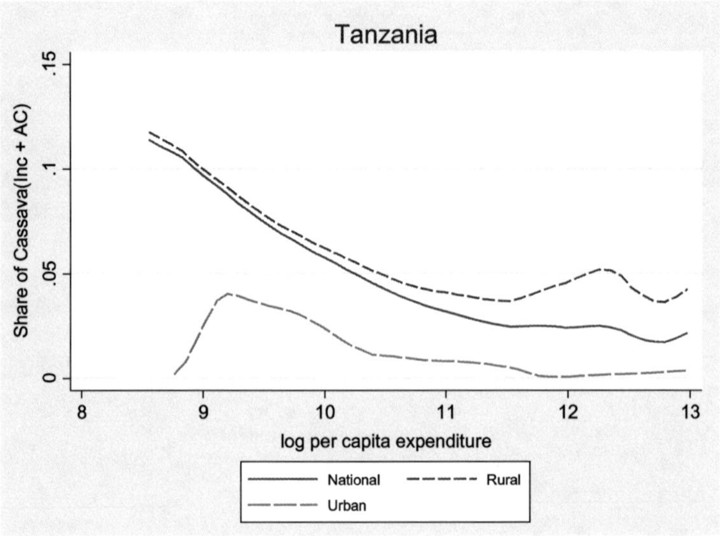

Fig. 3.9 Cassava income share across the income distribution. Source: Tanzania national panel survey (2008)

Nevertheless, at the bottom and the middle of the distribution, the income share of cassava is comparable to the budget share (although a bit smaller). Comparing regions, the curve for rural areas lies above the curve for urban areas, which implies that welfare effects in rural areas will tend to be larger than in urban areas.

We end this section with a quick look to some measures of infrastructure, household constraints and institution access. More concretely, based on the information from the household surveys, we look at education and services. According to this information, Tanzania has a low level of development in education. In Tanzania, 59.7 % of the population has no formal education, 25.6 % of population has complete primary education and only 0.5 % has complete secondary education. We also find that educational achievements are higher in urban than in rural areas. In urban regions 49 % of the population does not have formal education, but in rural regions the percentage rises to 62.8 %. Educational constraints, with known implications for agricultural productivity and labor opportunities, seem to be binding. Another indicator that provides information about household constraints and about institutional and infrastructure access is the average distance to important centers such as district head quarter, school, market milling machine, bank, police station, etc. In general, the data show that the population of Tanzania faces large distances to most of the important centers. For example, the average distance to district head quarter is 36.9 km, and to the government hospital is 26.6 km. When comparing regions, distance seems to be greater in rural areas, making these constraints even more binding.

3.2 Food and Cash Crops in Tanzania

Tanzania has excellent potential for agriculture-led economic growth. The country has abundant land and water resources, motivated agricultural entrepreneurs, and access to international markets through the major port of Dar es Salaam. Since 1985, the country's overall agricultural GDP has grown at an average annual rate of 3.3 % while the country's main food crops have been growing at 3.5 % annually and its export crops at 5.4 % annually. Further, the share of GDP provided by the agricultural sector, which provides labor for 75 % of the workforce, has been shrinking, from more than 30 % in 2000 to about 24 % in 2010, and is forecasted to decline further, to about 18 %, by 2025 (UN Food and Agriculture Organization [FAO], Monitoring and Analyzing Food and Agricultural Policies [MAFAP] program, 2013). Agricultural productivity, which grew by 20 % over the past decade and is measured here by value added per agricultural worker, has been one of the main drivers of this trend. The reintroduction of programs such as fertilizer subsidies through the National Agricultural Input Voucher Scheme (NAIVS), which increased government involvement in the sector, is one factor behind this productivity growth (Baltzer and Hansen 2011). In fact, this program was cancelled in 2014 as Tanzania experienced a production surplus which led to a decrease in commodity prices at the national level (ACET 2014). The major staples that are produced in Tanzania include maize, sorghum, millet, rice, wheat, pulses, cassava, potatoes, bananas and plantains with the bulk of the country's export crops being composed of coffee, cotton, cashew nut, tobacco, sisal, pyrethrum, tea, cloves, horticultural crops, oil seeds, spices and flowers. Smallholder farmers in Tanzania, who are the dominant leaders in the sector as a whole, support average farm sizes of between 0.9 and 3 ha and cultivate 5.1 million hectares annually, of which 85 % is food crops. Today, women compose the primary source of the agricultural labor force in the country, however as a result of the application of poor technology and a dependence on unreliable weather conditions, both the labor force and land productivity of the sector have begun to act as a limitation to the sector. The significance of the agricultural sector in terms of potential economic growth and poverty reduction in the country has been recognized by the government, which has also recognized the role that outside factors including infrastructure, rural financial services, land ownership and good governance have played and continue to play in the development of the sector.

In terms of policy, the National Agriculture Policy of 2013 (Tanzania 2013) is the most current agricultural policy in Tanzania. Through the NAP, the government made a commitment to "bring about green revolution that entails transformation of agriculture from subsistence farming towards commercialization and modernization through crop intensification, diversification, technology advancement, and infrastructural development." This agenda is in line with the National Strategy for Growth and Reduction of Poverty (NSGRP) and TDV 2025, which aims to raise the general standard of living of Tanzanians to the level of a medium-income country by 2025. The NAP 2013 is a comprehensive document covering every issue, from R&D and crop breeding to biofuel through agro-processing and a number of cross-cutting issues such as building resilience to climate change.

We have selected to analyze four crops for the case of Tanzania: cassava, maize, rice, and cotton. We want to see how price changes are sensitive to crop type. That is why we have chosen these four crops because they represent the cash/food crop dichotomy. We begin first with cassava. Cassava is an important subsistence food crop in Tanzania, especially in the semi-arid areas and sometimes considered as a famine reserve when cereals fail due to its drought tolerance. It is the second most important food crop after maize in terms of volume and per capita consumption, supporting livelihoods of about 37 % of farmers in rural areas. Tanzania is ranked the eighth largest producer of cassava in the world, and Africa's fifth largest. Cassava in Tanzania is widely grown in all farming systems due to its adaptability to various soils and agro-ecological conditions. The main cassava producing areas in Tanzania are the Lake Victoria zone, the Southern zone, the Eastern zone and the Southern Highlands. On average, Tanzania produces about five million tons of cassava per year and almost all is consumed domestically.

Cassava value chain faces a number of constraints. One of the main constraints is its negative reputation. In general, cassava is perceived as a food security or subsistence crop with little industrial or cash value. As a result, this negative image prevents it from being made a priority in agricultural policy deliberations. Until its image improves or is dealt with as in places such as Nigeria, it will continue to receive scant policy support in Tanzania. Second, while improved varieties of cassava are already being adopted in countries such as Nigeria, local cassava varieties with low root-yield potential continue to be favored over improved, high-yielding varieties, largely due to taste preferences. Improved varieties are more bitter than most of the local varieties, hence their low adoption rate. Last, the ultimate constraint facing the cassava value chain is the poor export market. In Tanzania, raw cassava is distributed through the consumption market, traditional or artisanal processors, and a small number of modern processors. There is apparently no export market for raw cassava in Tanzania.

Maize is the 5th agricultural commodity in Tanzania by value of production, accounting for 7.5 % of total production value. Moreover, it represents close to 5 % of total agricultural imports and is the main energy source in the diet accounting for 25 % of total caloric intake. Maize is grown across the whole of Tanzania but over half the national production comes from only a few regions, namely Iringa, Mbeya, Ruvuma and Rukwa. In terms of production, there has been an increase in production reaching 6.7 million tons in 2014 from 4.7 million tons in 2010. The increase in volume is partly due to fertilizer subsidy program which was introduced in 2008 through a voucher system. About 80 % of maize is produced by smallholder farmers. Between 65 and 80 % of all maize is consumed within the producing households leaving only 20–35 % for commercial trading within Tanzania.

Uncertainty is the key constraint facing the Tanzanian maize value chain. This include production, processing, market, storage, finance, and transport uncertainties. While some of these risks can be addressed through available risks mitigation tools, a number of market failure make these tools mostly inaccessible among smallholder farmers. This is the case for rainfall insurance where it was found that farmers willingness to pay for insurance is far below that actuarially fair rate (Sarris et al. 2006).

Rice is the second most important food crop in terms of number of households, area planted and production volume in Tanzania. Rice is among the major sources of employment, income and food security for Tanzania farming households. The leading regions in rice production are Shinyanga, Tabora, Mwanza, Mbeya, Rukwa and Morogoro. Rice is grown under three major ecosystems namely rain-fed lowland, upland rice and irrigated. About 74 % of total rice area is rain-fed lowland rice, 20 % is upland rice, and 6 % is irrigated. Although the rice statistics in Tanzania are erratic, it is estimated that, total available rice in the country, from imports and local production, usually ranges between 400,000 and 600,000 tons per annum with imports reaching up to 100,000 tons of rice per year.

High risk aversion among producers discourages adoption technologies such as improved seeds. In 2008, Tanzania's Agricultural Seed Agency (ASA) produced and marketed about 120,000 MT of improved rice seeds, but fewer than 10 % of rice farmers adopted them. One reason is that the seeds were not sufficiently popularized to reduce farmers' risk aversion; farmers were less willing to adopt because of limited knowledge. Lack of incentives for quality in processing is another constraint. Investment in new and modern milling equipment and

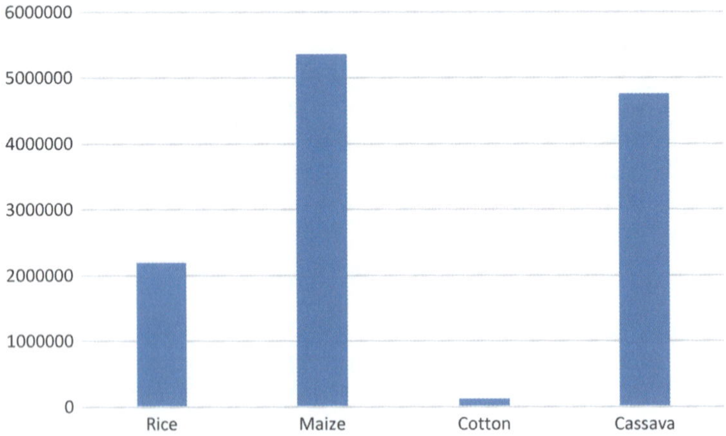

Fig. 3.10 Crops production in Tanzania (in tons, 2013). Source: FAOSTAT online

accompanying facilities, such as graders and destoners is hampered by a rice market that does not reward millers for the quality of rice they produce (ACET 2014).

Cotton is one of the major export crops for Tanzania, ranking second after coffee and grown by about 40 % of the total population in the country. The crop is annually grown on about 300,000–500,000 ha of land, which is equivalent to about 9 % of the total cultivated land in the country. More than 90 % of cotton is produced south of Lake Victoria in the Mwanza, Shinyanga, Mara, Tabora, Kigoma, and Singida regions, with Singida, Mwanza, and Shinyanga accounting for 80 % of it; the rest is produced in the Eastern part of the country. The production of seed cotton in Tanzania is characterized by smallholder farmers. Almost 90 % of cotton production comes from smallholders who cultivate cotton with average size of farms ranging from 1 to 2 ha. Due to small farm sizes and quantities of seed cotton produced, there is generally limited organization of production activities and logistics at the producer level. Part of the cotton lint produced in Tanzania is consumed locally by local textile mills. The textile industry is involved in dying, spinning, weaving, printing khanga and kitenge, bed sheets, garments, knitting, woven blankets and socks. Around 60 % of the production is exported of which the bulk consist mainly of cotton fiber (Figs. 3.10 and 3.11).

In terms of constraints it is important to first highlight that genetically modified organisms (GMOs) are illegal in Tanzania. As a result, (Bt) cotton, a genetically modified cotton widely adopted in West Africa, containing a protein that produces a natural insecticide, is not accepted by the local legislature, in line with Tanzania's fear of adopting GMOs in its agricultural sector. This precludes the sector from increasing productivity by lowering its output's vulnerability to pests. In terms of processing, the value chain is plagued by high proportion of old equipment. About 20 % of the country's spinning sector capacity is under 10 years old, nearly half the global average, and some mills have not been upgraded in over 40 years. Only about 8 % of mill spindles in Tanzania are less than 10 years old (Salm et al. 2011). On the

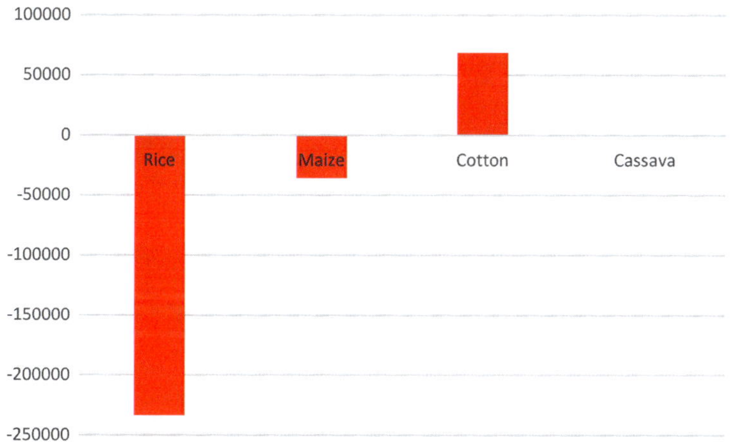

Fig. 3.11 Crops net export in Tanzania (in tons, 2013). Source: FAOSTAT online

political economy, there is a lack of trust between cotton farmers and community aggregators representing ginneries. An unfortunate new trend has arisen in the cotton value chain, in which some stakeholders deliberately contaminate cotton seed with trash and foreign matter in order to raise the weight of their production and thus their profit. Farmers and buying agents have blamed each other for these acts. The consequences are the low quality of Tanzania cotton (ACET 2014).

Table 3.3 below displays the market shares for processors of cassava, cotton, maize, and rice in Tanzania. Cassava is traded mostly informally. Of the formal processors, Ukaya process 20 tons per week or 34.48 % of the market followed by Mvinjeni with 25.86 % formal market share. The farm gate price for cassava is freely determined by the market forces. On the other hand, the government pre-determines the minimum cotton price and subsidizes it to insure higher farmers' welfare. The largest cotton processor in Tanzania, Kahama Cotton Company, controls around 8.35 % of the market followed by S&C Ginning Co with 8.07 % and Khahama Oil Mill and Afrisian Milling with 7 % each. There is a large number of maize processers in Tanzania processing about 6.7 million tons of maize per year. While the majority of maize is processed by a large number of very small local millers, there are some industrial players including Agro Processing & Allied Products Ltd that controls about 5.7 % of the quantity of milled maize, followed by the National Milling Corporation that controls 3.9 % of the market share. The maize market is a free market where the forces in the market determine the price at which it is traded. There is no government intervention in price setting since the liberalization of the market in 1996. However, the price of maize is determined by a lot of factors such as transportation costs which also determines the marketing margins of maize. There are large and volatile price differences across regions and farms in Tanzania, with the margin between retail and wholesale prices of maize from Iringa (one of the main suppliers of maize) to Dar es Salaam being consistently positive and increasing. Rice marketing in Tanzania is very fragmented such

Table 3.3 Market shares in Tanzania

Rice		Maize	
Company	Share (%)	Company	Share (%)
Southern	6.14	New Boogaloo Ltd	2.30
Export trading	5.57	Agro Processing And allied Products Ltd	5.74
Kilombero	7.71	Ben Es-Haq Ltds	2.30
Mtenda Kyela	21.57	Coast Miller Ltd	3.18
Othersmall scale miller	58.92	Kenmillers Ltd	1.44
		Mini Millers Ltd	0.19
		National Milling Corporation	3.98
		Said Salim Bakhresa and Co	1.91
		Remaining smaller millers	78.96
Cassava		**Cotton**	
Company	Share (%)	Company	Share (%)
Sululu	17.24	Kahama Cotton Company	8.35
Ukaya	34.48	S&C Ginning Co	8.07
Luna investment	22.42	Kahama Oil Mill	7.07
Mvinjeni	25.86	Afrisian Ginning	7.05
		Olam Ltd	6.92
		Gaki Investment	6.02
		Fresho Investment	4.80
		NidaTextile Mill	4.08
		JamboOil Mill	3.88
		NGS Investment Ltd	2.93
		Biore	2.24
		Others	38.59

Source: Stakeholders interviews

that over 90 % of the production is controlled by small scale traders. The remaining share of the rice market is shared among four players, including the Southern Highland Company, Export Trading, Kilombero Plantation limited, and Mtenda Kyela Rice Company. Of these major players, the first three are plantations and the fourth is a large scale trader who has a purchasing contract with about 15,000 farmers. Since the liberalization of the market, the rice market is a free market with no government intervention and marketing boards. Both the producer and wholesale prices for paddy in Tanzania is determined by the forces in the market as well as the international price of the commodity.

3.3 Simulation Results

In this section, we use the model to perform various simulations. These simulations are in fact comparative static results that stem from the model presented in Chap. 2. Among the parameters of the model, we consider two sets of exercises. Following

Porto et al. (2011), we shock the market structure of the supply chain. To this end, we consider (arbitrary) changes in the number of firms and in their market shares to capture both increases and decreases in the extent of competition in the supply chain. We study the cases of Leader split, Leaders merge, Exit of the largest firm, equal market shares, and a limit case of perfect competition. We also consider comparative static results from changes in key parameters affecting the production decision of the farmers. We explore (arbitrary) changes international prices, costs of production, endowments, risk and food security aversion. We are interested in price changes of the agricultural goods produced in Tanzania. The ultimate goal of these simulations is to feed the results to the household survey data to assess the welfare and poverty impacts.

We investigate four case studies: cotton, cassava (exportables), rice and maize (importables). Given the complexity of the scenarios, we simplify the analysis by working with a sequence of partial equilibrium models so that each case study is dealt with separately. This just means that, in the case of cotton for instance, we keep all the other markets unmodeled.

In what follows, we describe in detail the results for the case of cotton. We later list the major findings for the other case studies, highlighting differences and specific results. Cotton is a cash crop exported by Tanzania and we thus use the cash crop export model. The price changes from the simulation results are presented in Table 3.4. The first row shows the impacts of changes in competition. As expected, increases in competition raise farmgate cotton prices, while decreases in the extent of competition reduce prices. In the case of Leaders merge and Exit of the largest, prices decline by 3.44 and 3.52 % respectively. In contrast, the splits of

Table 3.4 Farmgate price simulation results for cotton

	Baseline	Leader split	Leaders merge	Exit of largest	Equal marquet shares	Perfect competition
Competition policy	0.00	2.56	−3.44	−3.52	1.55	18.17
Increase of 10 % in						
International price	18.82	22.06	14.32	14.20	20.40	40.90
Marginal cost of producing cash crop	0.61	3.11	−2.27	−2.40	2.22	18.18
Fixed cost of producing cash crop	0.56	3.10	−2.40	−2.51	2.13	18.18
Household resources (endowment)	−2.21	0.59	−5.82	−5.97	−0.78	18.18
Risk and food security Paremeter	2.81	4.88	0.24	0.15	4.28	18.18
Food crop price	0.40	2.81	−2.63	−2.72	1.94	18.18
Marginal cost of producing Food Crop	−0.17	2.45	−3.67	−3.77	1.41	18.18
Non-farmer demand	–	–	–	–	–	–

Source: Simulation results from the model of Chap. 2

the leader would increase prices by 2.56 %, and move to an oligopsony with equal market shares, by 1.55 %, and a move to a limit case of perfect competition, by 18.17 %. Note that all these effects are moderate, except the extreme move to competition.

The role of household constraints is explored in column 1 of Table 3.4, starting in row 2. This is the baseline model, where the structure of the market chain is not shocked. Different rows correspond to different comparative static results. International prices have large impacts on farmgate prices. In the margin, after a price increase of 10 %, for example, farmgate prices would increase by almost 19 %. This implies a pass-through rate of 1.9. This is consistent with findings in Porto et al. (2011), on which our model builds, but it is a large elasticity.

Increases in the marginal cost and in the fixed cost of producing cotton lead to increases in farmgate prices (rows 3 and 4). This is a very intuitive result because higher costs imply a shift up in the farm aggregate cotton supply and a consequent increase in equilibrium prices. Note that the response of prices is, however, cushioned to a very large extent by the market structure: cost increases of 10 % lead to price changes of roughly half of a percentage point. The increase in the endowment has the opposite effect (row 5). An exogenous increase in endowment means more resources for the farmers. They can thus more easily satisfy any food security needs and leave more resources available for the production of the cash crop. This implies an increase in farm cotton supply and a lower equilibrium price. Note, once again, that the impacts are cushioned by imperfect competition among processors. The implications of these results are straightforward. Cost reductions (increases) in cotton production benefit (hurt) farmers but the general equilibrium effects via prices may hurt (benefit) them. Nevertheless, the price effects are scaled down by the competition between exporters so that the direct effect appears to dominate.

The model also predicts that increases in household risks that lead to higher demands for food security positively affect equilibrium cash crop prices (row 6). This is an interesting result. The intuition works along the same lines as before. Imagine shocks to farmers that induce them to want to better cover their food needs via subsistence activities. This could be caused by more erratic food market conditions, a higher health risk for productive household members, and so on. In this scenario, households react by retracting to auto-consumption and by allocating more resources to auto-consumption and thus lower resources to cotton. In the end, cotton farm supply is lower, and cotton prices may increase in equilibrium. This can benefit cotton farmers. As we will show below, this result suggests that negative and unwanted shocks to food producers (in rural areas, for example) may end up benefiting cotton producers. This may exacerbate inequality between farmers and increase relative poverty impacts, for example.

An important element in our model is that it allows us to explore, at least to some extent, the spillovers and interrelationships between cash crop production and food markets. In the cotton export model, farmers take the prices of competing marketable foods as given, but the level of these prices clearly affects production and consumption decision. Similarly, the marginal cost of producing food can also affect cotton production choices. In our simulations, we find that increases in the prices of competing food crop prices causes an increase in cotton prices (row 7). Alternatively, an increase in the cost of producing those goods can lead to general equilibrium declines in cotton prices (row 8). Consider an increase in the price of marketable food. This induces farmers to produce more food and less cotton and the price of cotton increases as a result. The opposite would happen if the cost of producing the marketable food increases. It is important to emphasize these results. They highlight the role of stressing the feedback effects between food production and cash crop export production. These feedbacks are seldom studied in the literature but our model shows they can be sizeable.

To end, we examine complementarities between shocks to the structure of competition among exporters and shocks to household constraints. The idea is to uncover potential synergies between different types of policies or shocks. For instance, an increase in competition among exporters brings farmgate cotton prices up. The same happens when the international price increases. Complementarities would occur if the change in farmgate price due to the increase in competition is boosted by a concurrent increase in international prices (net of the direct effect of these higher prices). It is not easy to establish these complementarities quantitatively. Our approach here is to simulate the impacts of the joint shocks and to compare these numbers with the sum of the impacts of each individual shock. Table 3.4 reports the joint effect. The sum of the separate effects can be easily calculated from the competition policies shocks (row 1) with the baseline complementary policy results (column 1).

Our model features complementarities and substitutabilities. It is difficult to generalize the results, however. Complementarities show up when the joint effect is larger than the sum of the separate effects. Consider, for instance, the case of Equal market shares and higher marginal costs. The joint effect of those two shocks would be an increase in farmgate prices of 2.22 %. Instead, the sum of the separate effects is smaller, 2.16 %. In this case, the complementarity exists but is small (equivalent to roughly 3 % of the joint effect). In other cases, the complementarity is much larger. The intuition is that the increase in competition causes prices to increase and this increase is larger if, concurrently, the costs of producing cotton are larger.

Consider now the case of Leader Splits together with an increase in food security and household risk. The result of the joint shock would be an increase in cotton farmgate prices of 4.88 %. Instead, the sum of the separate shocks would bring prices up by 5.37 %. This is a "substitutability" effects that implies a difference of 10 %, approximately. In this case, the increase in food security risks induce farmers to reallocate resources out of cotton and into food, thus reducing cotton supply and increasing cotton prices. When this happens in the presence of more competition, which in itself implies higher prices, the reallocation of resources is ameliorated and the price increase is therefore smaller.

As we mentioned above, while the model delivers complementarities and substitutabilities, it is difficult to generalize and to find clear patterns in the results. Sometimes, shocks and policies go in the same direction, sometimes they oppose each other. Sometimes the joint effects are big, sometimes they are small. The

Table 3.5 Farmgate price simulation results for cassava

	Baseline	Leader split	Leaders merge	Exit of largest	Equal marquet shares	Perfect competition
Competition policy	0.00	0.06	−0.06	−0.06	0.00	1.28
Increase of 10 % in						
International price	11.59	11.69	11.48	11.48	11.59	13.78
Marginal cost of producing cash crop	−0.03	0.03	−0.10	−0.10	−0.03	1.28
Fixed cost of producing cash crop	−0.33	−0.26	−0.41	−0.41	−0.33	1.28
Household resources (endowment)	0.18	0.23	0.13	0.13	0.18	1.28
Risk and food security parameter	0.68	0.71	0.66	0.66	0.68	1.28
Cash crop price	0.83	0.85	0.81	0.81	0.83	1.28
Marginal cost of producing Food Crop	0.16	0.20	0.11	0.11	0.16	1.28
Non-farmer demand	0.07	0.13	0.01	0.01	0.07	1.28

Source: Simulation results from the model of Chap. 2

important lesson from these exercises, beyond the quantification of the special cases considered in the simulations, is that these complementarities exist and need to be taken seriously in the design of agricultural policies.

The other exportable crop that we study is cassava. This is considered as an exportable food crop in Tanzania. Results are reported in Table 3.5. Given the nature of competition in the supply chain, which is highly competitive in the baseline, changes in the structure of the market has little impact on cassava farmgate prices. This can be seen in the first row of Table 3.5. In general, the price changes are negligible. Even in the move to a limit case of perfect competition (thus eliminating any remaining imperfections in the cassava market), prices would increase by only 1.28 %. As in the case of cotton, changes in international prices have large impacts on cassava prices. This is roughly consistent with a setting where many traders compete with each other.

It is noteworthy that shocks to complementary factor seem to have small effects on cassava prices. We can divide these factors in two sets. On the one hand, there are factors that affect directly the production of cassava, such as household risks and cassava production costs. On the other hand, recall that our food export model includes a farm production decision that allows farmers to choose between cassava and a competing cash crop (e.g., cotton). This means we can look at feedbacks and spillovers from cash crop markets to food (exportable) markets.

In the case of cassava, we find that changes in production costs of the cash crop generate reduction in the price of cassava. This is because higher costs of producing cash crops induce a shift of resources out of the cash crop and into the competing food export crop, thus increase cassava supply. The magnitudes are, however, very small. The impact of changes in marginal costs is −0.03 % (row 3) and the impact of changes in fixed costs is −0.33 % (row 4). An increase in the price of the cash crop, in turn raises cassava prices (row 7) because it induces farms to produce more cash crop and supply less cassava.

Factors that affect cassava production directly also have small impacts. In row 8, for instance, a 10 % increase in the marginal cost of producing cassava raises cassava equilibrium prices by only 0.16 %. Similarly, an increase in household total resources does not affect prices much (row 5). Similarly, changes in household risks that raise auto-consumption have a positive effect on cassava prices. The magnitudes are small, but still larger than for other complementary shocks (row 6). For instance, in the baseline, the price change caused by an increase in household risk would be of 0.68 %. The best explanation we can provide for these findings is that while cassava is an important food crop in Tanzania, a lot of it is not channelled through the export market. As a result, changes in household constraints seem to have only small effects on the aggregate cassava supply. Combined with a competitive setting, the equilibrium price change is finally small.

We now turn to the case of rice and maize, two importable food commodities. Both are important crops, both in production and in consumption, and maize appears to be more relevant than rice (in terms of budget and incomes shares at least). We thus begin with the case of maize in Table 3.6. As shown above, the maize market is dominated by a large number of traders and intermediaries. Analytically, the market behaves very competitively and, as a result, the shocks to market structure imply negligible impacts on prices (row 1). Note that we are dealing with an importable commodity and thus the market is represented by an

Table 3.6 Farmgate price simulation results for maize

	Baseline	Leader split	Leaders merge	Exit of largest	Equal marquet shares	Perfect competition
Competition policy	0.00	−0.02	0.02	0.02	0.00	−0.44
Increase of 10 % in						
International price	8.43	8.44	8.42	8.42	8.43	8.65
Marginal cost of producing cash crop	−0.17	−0.18	−0.16	−0.16	−0.17	−0.44
Fixed cost of producing cash crop	−0.18	−0.19	−0.17	−0.17	−0.18	−0.44
Household resources (endowment)	0.24	0.21	0.27	0.27	0.24	−0.44
Risk and food security paremeter	0.38	0.35	0.42	0.42	0.38	−0.44
Cash crop price	0.84	0.79	0.91	0.91	0.84	−0.44
Marginal cost of producing food crop	0.15	0.12	0.18	0.18	0.15	−0.44
Non-farmer demand	0.03	0.01	0.05	0.05	0.03	−0.44

Source: Simulation results from the model of Chap. 2

oligopoly. In consequence, increases in competition should bring prices down (see for example, the price decline in the limit case of perfect competition).

Increases in international prices are transmitted to the local economy, in part due to the nature of competition (row 2). In general, complementary factors affecting household constraints and resources have also small impacts on prices. Only an increase in the cash crop price (of 10 %, row 8), which is in principle more profitable, elicits a supply response that increases maize prices by 0.84 %. These results imply that the maize market is generally inelastic to the shocks considered in our exercises. This is because, even though maize is an important food crop, the marketable fraction of production is typically small. Moreover, given the appropriate conditions, the model implies that the cash export crop is more profitable and, consequently, farmers prioritize resource reallocation to these crops.

An interesting novel results (albeit a weak one) that arise in the case of maize if the following. When the endowment is higher, the price of maize increases (slightly). A higher endowment allows households to produce more of all crops, including maize. Ceteris paribus, this should lead price to decline because of a larger supply. However, the price of maize increases. This could happen if the increase in household resources is such that cash crop production becomes, at the margin, profitable to a larger number of farmers and this creates incentives to move some resources out of maize and into the export cash crop. It is difficult to establish this result more generally, but it is another interesting finding that highlights feedback and spillovers across markets and household activities and decisions.

Results for rice are presented in Table 3.7. Overall, the simulations for rice resemble qualitatively the results for maize, although the magnitudes of the effects

Table 3.7 Farmgate price simulation results for rice

	Baseline	Leader split	Leaders merge	Exit of largest	Equal marquet shares	Perfect competition
Competition policy	0.00	−0.03	0.04	0.04	0.00	−0.85
Increase of 10 % in						
International price	8.84	8.81	8.87	8.87	8.84	8.24
Marginal cost of producing cash crop	−0.09	−0.12	−0.05	−0.05	−0.09	−0.85
Fixed cost of producing cash crop	−0.03	−0.06	0.02	0.02	−0.03	−0.85
Household resources (endowment)	−0.03	−0.06	0.02	0.02	−0.03	−0.85
Risk and food security paremeter	0.14	0.09	0.19	0.19	0.14	−0.85
Cash crop price	0.23	0.18	0.29	0.29	0.23	−0.85
Marginal cost of producing food crop	0.10	0.05	0.15	0.15	0.10	−0.85
Non-farmer demand	0.11	0.06	0.16	0.16	0.11	−0.85

Source: Simulation results from the model of Chap. 2

are somewhat smaller because rice is less relevant than maize. The rice market appears to be quite competitive and, consequently, changes in the nature of competition among importer trading bring only small price changes. The role of complementary factors is also weak. Apart from changes in border prices, which transmit to the local economy in a fashion similar to maize, all the other parameters of the model generate very small prices changes. The intuition of the maize case study applies here as well (competitive markets combined with a crop that farmers market only limitedly).

3.4 Welfare Simulations

We end our analysis with a discussion of the poverty impacts of the comparative static results presented above. Ultimately, we are interested in the role of the supply chain in agriculture on household well-being, on whether the poor are affected more or less than the non-poor, and on whether the complementarities between the structure of markets and household constraints can inform policy about ways to boost or ameliorate those poverty impacts. This is the goal of this section.

The analysis is done using standard techniques in the literature. We adopt the first order approximation analysis of Deaton (1989a, b; 1997). This implies we can approximate the impact of a price change using income shares and budget shares as measures of exposure. The first order approximation works well if the price changes are small and if there are limited supply and consumption responses. It is, in general, a very powerful and useful tool to evaluate the welfare effects of price changes.

The welfare impacts of the price changes are reported in Tables 3.8, 3.9, 3.10 and 3.11 for the cases of cotton, cassava, maize and rice. We show the impacts of shocks to the market structure. To illustrate the complementarities, we show results for a combination of shocks to market structure and international prices (we comment on the results for other complementarities at the end). We also report average results for the total population, the poor, and the non-poor. In the case of cotton, we report separate results for cotton producers.

Some regularities can be detected in the simulation results. Increased competition and complementary policies in cotton show positive welfare impacts across households. The impacts are obviously larger for cotton producers. Competition among exporters in a cash export crop implies higher farm-gate prices and, consequently, higher farm income from cotton production. Since raw cotton is only produced and not consumed directly by the households, real farm income is in the end higher. Even though there is net production of cassava, competition and higher prices create (small) welfare losses because of the distribution of consumption shares among both producers and consumers. In addition, higher maize and rice prices (due to lower competition in the supply chain) create welfare losses because these are staple crops.

Table 3.8 Cotton price changes and household welfare

	Baseline	Leader split	Leaders merge	Exit of largest	Equal marquet shares	Perfect competition
Total						
Competition policy	0.00	0.02	−0.03	−0.03	0.01	0.16
International price	0.17	0.20	0.13	0.13	0.18	0.37
Poor						
Competition policy	0.00	0.01	−0.02	−0.02	0.01	0.08
International price	0.08	0.10	0.06	0.06	0.09	0.18
Non poor						
Competition policy	0.00	0.04	−0.05	−0.06	0.02	0.29
International price	0.30	0.35	0.23	0.23	0.32	0.65
Producers						
Competition policy	0.00	0.44	−0.59	−0.60	0.26	3.11
International price	3.22	3.77	2.45	2.43	3.49	6.99

Note: First order impact on household welfare

Table 3.9 Cassava price changes and household welfare

% Variariation in utility	Baseline	Leader split	Leaders merge	Exit of largest	Equal marquet shares	Perfect competition
Total						
Competition policy	0.00	0.00	0.00	0.00	0.00	−0.01
International price	−0.11	−0.12	−0.11	−0.11	−0.11	−0.14
Poor						
Competition policy	0.00	0.00	0.00	0.00	0.00	−0.01
International price	−0.06	−0.06	−0.06	−0.06	−0.06	−0.07
Non poor						
Competition policy	0.00	0.00	0.00	0.00	0.00	−0.02
International price	−0.19	−0.19	−0.18	−0.18	−0.19	−0.22

Note: First order impact on household welfare

Table 3.10 Maize price changes and household welfare

	Baseline	Leader split	Leaders merge	Exit of largest	Equal marquet shares	Perfect competition
Total						
Competition policy	0.00	0.00	0.00	0.00	0.00	0.02
International price	−0.46	−0.46	−0.46	−0.46	−0.46	−0.48
Poor						
Competition policy	0.00	0.00	0.00	0.00	0.00	0.02
International price	−0.30	−0.30	−0.30	−0.30	−0.30	−0.31
Non poor						
Competition policy	0.00	0.00	0.00	0.00	0.00	0.04
International price	−0.68	−0.68	−0.68	−0.68	−0.68	−0.70

Note: First order impact on household welfare

Table 3.11 Rice price changes and household welfare

	Baseline	Leader split	Leaders merge	Exit of largest	Equal marquet shares	Perfect competition
Total						
Competition policy	0.00	0.00	0.00	0.00	0.00	0.02
International price	−0.22	−0.22	−0.22	−0.22	−0.22	−0.21
Poor						
Competition policy	0.00	0.00	0.00	0.00	0.00	0.00
International price	−0.05	−0.05	−0.05	−0.05	−0.05	−0.04
Non poor						
Competition policy	0.00	0.00	0.00	0.00	0.00	0.01
International price	−0.14	−0.14	−0.14	−0.14	−0.14	−0.13

Note: First order impact on household welfare

To a large extent, the welfare impacts are small for all groups of households. For most crops, shocks, and affected population, the welfare impacts of the proposed simulations are <1 % of total household expenditures. The only exception is the impact on cotton producers where some sizeable impacts can be established.

These results are expected, given the nature of the exercise considered here, and they are also comparable to the literature on the topic (see the review in Lederman and Porto 2015). There are various elements that need to be taken into account. First, the income shares and budget shares used in the first order approximation are typically small (recall the household survey analysis of Sect. 3.1). Some crops are relevant separately on both the production side and on the consumption side. But a price change affects households as consumers and as producers, and thus the net effect tends to be small in general. Second, in the crops considered here, the market was already characterized by some degree of competition, thus leaving small room for sizeable price changes. The combination of small price changes with small net benefit ratios (Deaton 1997) implies small impacts.

The fact that the impacts are typically small does not mean they are not important. As we argued above, small results are expected in this literature. They are expected given the context (household survey data and baseline market structure) but are reasonable. We are just assessing the short-run impacts of price changes caused by changes in exporters' market power and the combination with complementary factors. It is important to note that the complementary factors have an independent effect on household welfare that we are not attempting the measure here. If, for instance, the cost of crop production declines due to improvement in infrastructure, access to cheaper and better inputs, access to knowledge or credit, etc., there will be a direct impact on welfare and an indirect one via the combination with changes in market structure. Here, we are measuring this additional impact only. It turns out that these additional impacts are small but, since they do not carry additional costs (for example fiscal costs if the complementarities are funded by the government), they only generate benefits.

References

African Center for Economic Transformation (ACET) (2014) Tanzania country report. Sustainable rural development project, Accra, Ghana

Baltzer KT, Hansen H (2011) Agricultural input subsidies in Sub Saharan Africa. Ministry of Foreign Affairs of Denmark, Danida, pp 8–36

Deaton A (1989a) Rice prices and income distribution in Thailand: a non-parametric analysis. Econ J 99(395):1–37

Deaton A (1989b) Household survey data and pricing policies in developing countries. World Bank Econ Rev 3:183–210

Deaton A (1997) The analysis of household surveys: a microeconometric approach to development policy. Johns Hopkins University Press for the World Bank, Baltimore, MD

Lederman D, Porto G (2015) The price is not always right: on the impacts of (Commodity) prices on households (and Countries). World Bank Res Obs 31(1):168–197

Porto G, Depetris Chauvin N, Olarreaga M (2011) Supply chains in export agriculture, competition, and poverty in Sub-Saharan Africa. CEPR Press, London

Salm A, Dinsdale P, MacDonald D, Martelli C, Hill K, Kabissa J (2011) Tanzania textiles and garment development strategy. Interim report for the ministry of industry and trade. A study funded by Tanzania Gatsby Trust

Sarris A, Karfakis P, Christiaensen L (2006) Producer demand and welfare benefits of rainfall insurance in Tanzania. FAO Commodity Trade Policy Res Work Pap (18):1–33

Tanzania (2013) National Agriculture Policy. Ministry of Agriculture, Food Security and Cooperatives, October 2013

Chapter 4
The Case of Malawi

In this chapter we cover the case of Malawi. The household survey data is described in Sect. 4.1. We summarize the food and cash crops institutional arrangements in Sect. 4.2. We present the farmgate price simulation results from the model in Sect. 4.3 and the household welfare simulations in Sect. 4.4.

4.1 The Household Survey Data

For Malawi, the household data comes from the Second Integrated Household Survey, 2004/2005. The dataset contains information on more than 11,000 households. The overwhelming majority of the households are rural (87 %), being the remaining 13 % urban. When it comes to age, the population is very young, with 46 % being 14 years old or less, and only 3.7 % above 65. Gender wise, the population is fairly balanced: 50.7 % female and 49.3 % male. However, the balance is broken when it comes to household heads: only 23 % of them are female, 24 % in rural areas and 15 % in urban ones. Household sizes tend to be big: 5.8 people on average, of which 33 % are under 18 years of age. Household sizes hardly differ between rural and urban areas (5.89 and 5.68 respectively).

Figure 4.1 shows the distribution of income. The graph shows the estimated density function of the logarithm of household per capita expenditure at the national level and for urban and rural regions separately. As usual, the distribution for urban households lies to the right of the rural distribution, meaning urban households have a higher expenditure level than rural ones on average. Since the rural sample is almost seven times larger than the urban one, the national distribution is almost identical to the rural one.

We now describe the sources of income and the patterns of consumption across households. In order to do this we turn to Table 4.1, where we report consumption patterns both for urban and rural regions, both cash expenditures and the value of

N. Depetris Chauvin et al., *Agricultural Supply Chains, Growth and Poverty in Sub-Saharan Africa*, Advances in African Economic, Social and Political Development, DOI 10.1007/978-3-662-53858-6_4

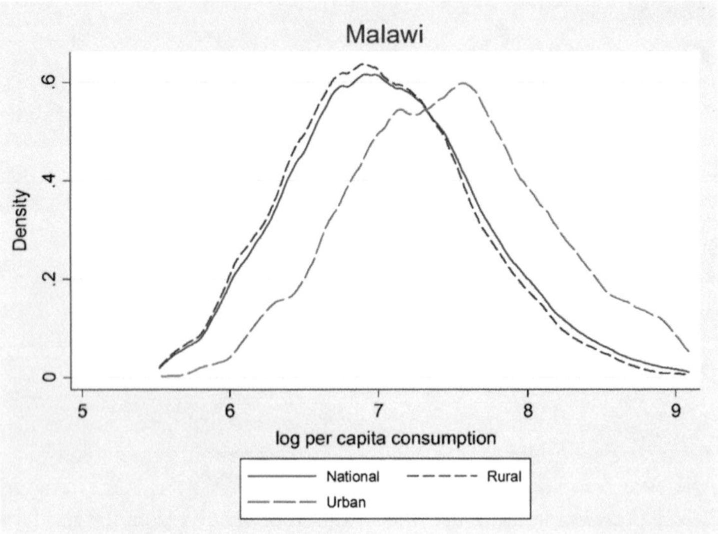

Fig. 4.1 The distribution of income density of (log) per capita household expenditure. Source: Malawi Second Integrated Household Survey (2004/2005)

auto-consumption. The latter is much higher than the former in rural areas, which is to be expected. We focus on the share of food consumption in the households' budgets, which is similar in rural and urban households, 58.8 % and 53.4 % respectively. This is coherent with our previous data on the distribution of expenditures presented above and Engel's law: urban households are richer than rural households, and they therefore spend a smaller share of their budget in food. The most relevant crop we have in the data is maize, amounting to 6.4 % of total expenditures, 6.3 % for rural and 8.9 % for urban household. Cash crops are insignificant in consumption, as expected.

In Table 4.2, we focus on the income side. As it usually happens, auto-consumption is much larger in rural households than in urban households (48.5 versus 12.5 %). Disaggregating the data reveals that maize is the most significant crop, with 23 % of total income for rural and 7.4 % of total income for urban households, being tobacco a far second 4.6 % for rural and 1 % for urban households respectively.

In order to be able to analyze the welfare effects of changes in prices of these commodities, we must first describe the patterns of income and expenditure sources along the income distribution. We will take the log of per capita household expenditure (log pce) as a measure of income and we plot estimates of non-parametric regression of income and budget shares on log pce.

Figures 4.2 and 4.3 refer to the average share of food expenditure and the average share of food income, both including cash expenditures as well as auto-consumption. In both cases, food share slopes downward, except for lower income levels. The fall is sharp; from 60 % of total consumption/income to 40 %. In the

Table 4.1 Budget shares

Malawi	Total	Rural	Urban
Total consumption per capita	100.0	100.0	100.0
Expenditures	68.8	65.7	93.8
Food	27.2	24.6	49.1
Manufactures	28.8	28.8	28.3
Services	8.0	7.4	12.8
Others	4.8	5.0	3.7
Auto-consumption	31.2	34.3	6.2
Auto-consumption food	31.2	34.3	6.2
Auto-consumption others	0.0	0.0	0.0
Total food consumption	58.5	58.8	55.3
Total crops	29.5	30.3	22.5
Maize	20.9	21.9	13.0
Rice	1.1	1.0	2.2
Poultry	1.7	1.7	2.1
Livestock	1.8	1.6	3.0
Cassava	1.4	1.4	0.9
Sorghum	0.2	0.2	0.0
Millet	0.1	0.1	0.0
Cowpea	1.2	1.3	0.3
Yam	0.8	0.8	0.7
Cotton	0.0	0.0	0.0
Tobacco	0.2	0.2	0.2

Source: Malawi Second Integrated Household Survey (2004/2005)

graph involving income, the difference arises simply because rural households are food producers. From these two figures we can draw some preliminary conclusions about the welfare effects of food price changes. On the consumption side, price declines will improve poor households' welfare more than wealthy households', and rural rather than urban households'. On the production side, however, the opposite is true: food price declines will hurt rural significantly more than urban ones, and poor rather than richer households. The consumption effect seems to prevail for urban households, while for rural ones we must analyze the data more closely.

That is why we move on to study specific crops (Figs. 4.4 and 4.5). Since the only relevant food crop is maize, it will be the only one included in this preliminary analysis. Maize shows the same patterns as the more general food income/expenditure graphs, which is to be expected, given that maize dominates the rest of the food crops in importance, as we saw in Tables 4.1 and 4.2. At each income level, urban producers consume less maize than rural households, which is probably given by different relative prices of maize between the regions. On the income side, urban households' income depends significantly less on maize production. Both the share

Table 4.2 Income shares

Malawi	Total	Rural	Urban
Total income per capita	100.0	100.0	100.0
Incomes	55.4	51.5	87.5
Food (agriculture)	11.9	13.0	3.0
Wage	14.1	10.4	45.1
Enterprises	14.3	12.9	25.9
Transfers	15.1	15.2	13.5
Auto-consumption	44.6	48.5	12.5
Auto-consumption food	44.6	48.5	12.5
Auto-consumption others	0.0	0.0	0.0
Total food income and AC	56.5	61.5	15.5
Total naps	35.4	38.5	9.8
Maize	21.3	23.0	7.4
Rice	1.0	1.1	0.0
Poultry	2.2	2.5	0.5
Livestock	1.8	1.9	0.3
Cassava	1.5	1.6	0.3
Sorghum	0.3	0.3	0.0
Millet	0.1	0.1	0.0
Cowpea	0.6	1.3	0.1
Yam	0.9	1.0	0.2
Cotton	0.5	0.5	0.0
Tobacco	4.2	4.6	1.0

Source: Malawi Second Integrated Household Survey (2004/2005)

of maize consumption and income is declining in income for urban and rural households.

We also show non-parametric regressions of cash crops. The main difference between cash and food crops is that households don't consume the former, therefore we only show the income side in Figs. 4.6 and 4.7. Figure 4.6 refers to the cash crop cotton. We can see that it has the same pattern as maize, declining in income, but with a much lower share in total income, zero particularly for urban households. Figure 4.7 shows the same for Tobacco. Again, shares are smaller for urban households, but the share in income is increasing in income for rural households, therefore, tobacco production corresponds to wealthier rural households.

4.2 Food and Cash Crops in Malawi

Malawi enjoys a sub-tropical climate and the weather is usually quite predictable. As a result, agriculture is naturally a key sector of the economy. Representing 33 % of GDP, the agricultural sector provided employment for 51.4 % of the active population in 2012 compared to 62.3 % in 2000 (McMillan and Rodrik 2011). The

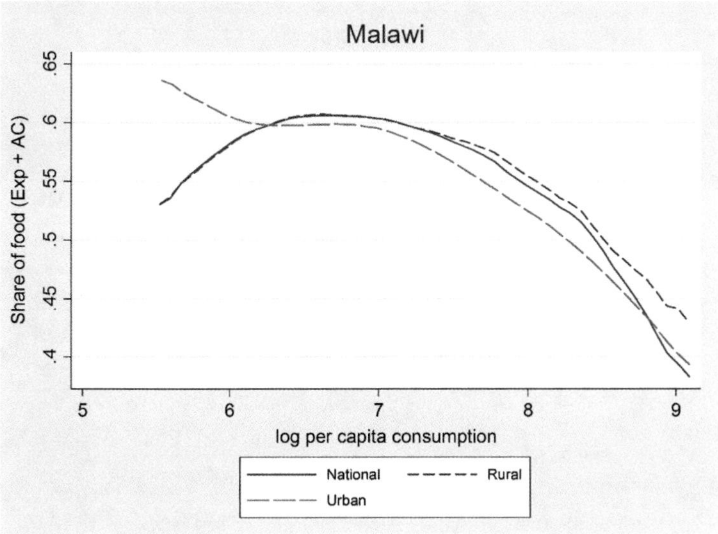

Fig. 4.2 Total food budget share across the income distribution. Source: Malawi Second Integrated Household Survey (2004/2005)

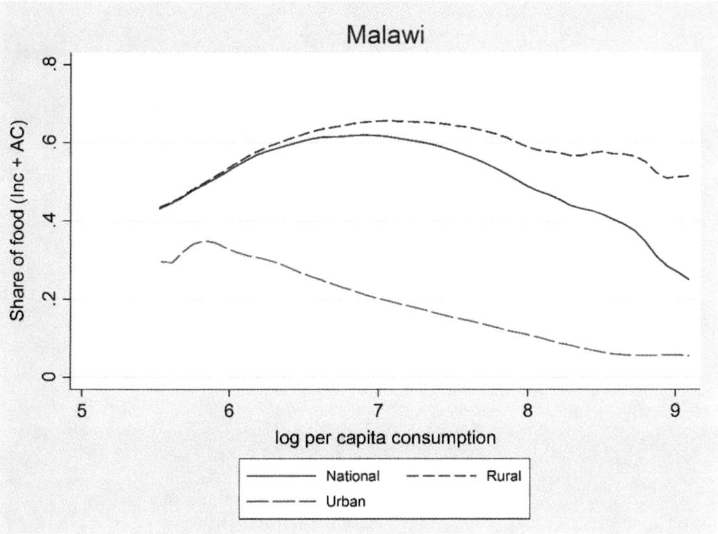

Fig. 4.3 Total food income share across the income distribution. Source: Malawi Second Integrated Household Survey (2004/2005)

major staples that are produced in Malawi in terms of value include cassava, potatoes, maize, banana, and dry beans with the bulk of the country's export crops being composed of wheat, tobacco, and in recent years, maize. Smallholder

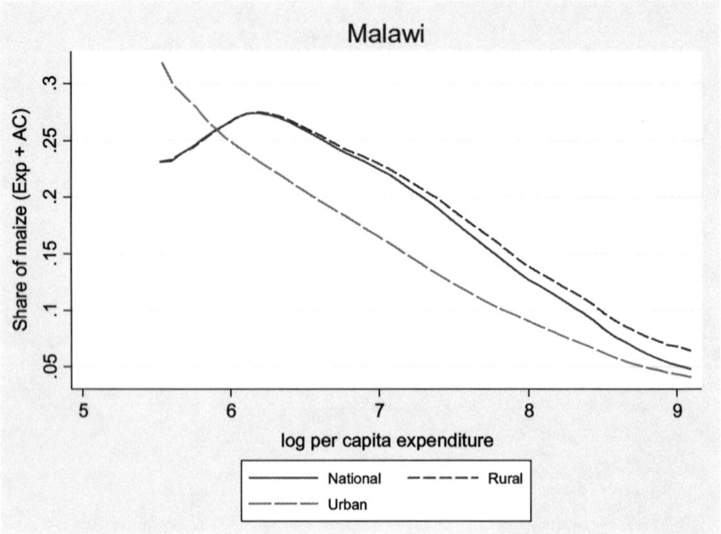

Fig. 4.4 Maize budget share across the income distribution. Source: Malawi Second Integrated Household Survey (2004/2005)

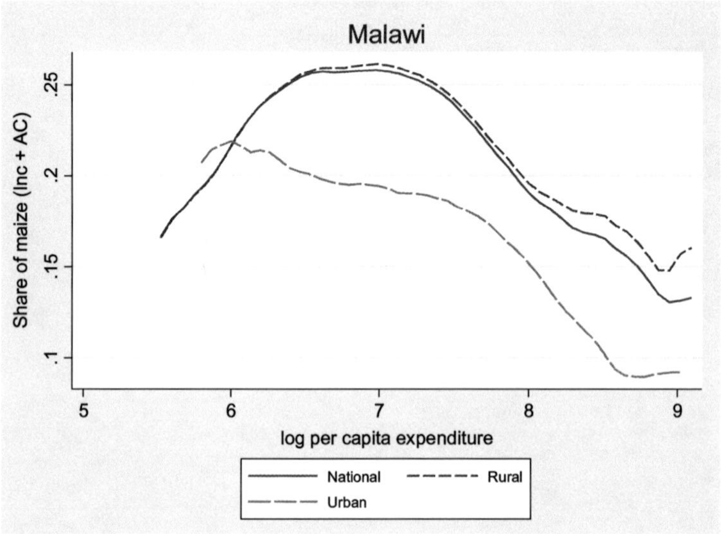

Fig. 4.5 Maize income share across the income distribution. Source: Malawi Second Integrated Household Survey (2004/2005)

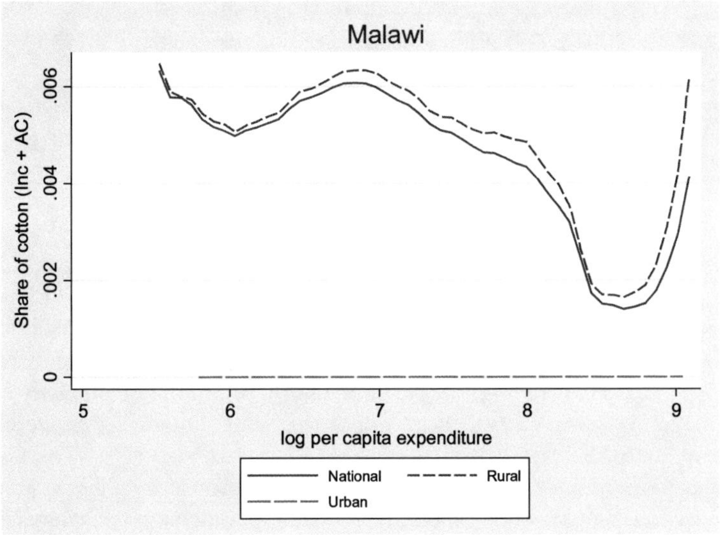

Fig. 4.6 Cotton income share across the income distribution. Source: Malawi Second Integrated Household Survey (2004/2005)

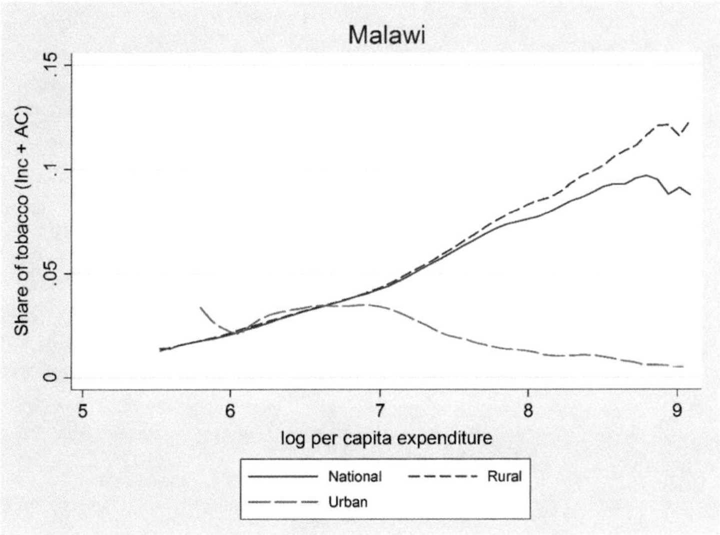

Fig. 4.7 Tobacco income share across the income distribution. Source: Malawi Second Integrated Household Survey (2004/2005)

farmers in Malawi, who are the dominant leaders in the sector as a whole, support average farm sizes of 0.72 ha and cultivate a total of 1.58 million hectares annually, of which 48 % is allocated to cereals (Kharas et al. 2015). Agricultural growth accelerated from around 4 % in 2004/2005 to around 14 % in 2006/2007 and to around 13 % in 2008/2009. Within the same period the economy grew by 8.6 % in 2007, 9.7 % in 2008, and 7.6 % in 2009.

The main policy document that currently guides Malawi's government actions in agriculture is the National Agriculture Policy (NAP). It is themed "Promoting agricultural productivity for national food security and economic growth and development through value chain development". Endorsed after a series of stakeholder consultations that started in 2009, the NAP was adopted 5 years later in 2014. Malawi is known for being a precursor to the now widely adopted fertilizer subsidy program. While most African countries were adopting Washington consensus policies that basically recommended government hands off the agricultural sector, Malawi went against the tide and introduced the farm input subsidy program (FISP) that gave access to fertilizers to more than 1.5 million farmers. Introduced in 2004/2005, FISP turned a net maize importer country into a net maize exporter country by 2006/2007.

We focus our crop analysis on maize, cotton, and tobacco (Figs. 4.8 and 4.9). Maize is by far the most important crop produced in Malawi. It constitutes 89 % of all cereals produced in Malawi. In 2014, total maize production was 3.9 million metric tons, or 56 % more than the 2000 production. In the period 1990–2010, yields have increased from 0.99 to 2.3 t/ha (Derlagen 2012). When taking a closer look, this strong increase in production and yield has largely been realized since 2006 and is generally considered to be related to the FISP that has been in effect since the 2005/2006 harvest season, as well as favorable weather conditions (Derlagen 2012). The central region of Malawi is the main production area; it represents 59 % of total production. The southern region counts with 45 % of the country's population but only 17 % of total maize production in 2007/2008. As indicated by FEWSNET (2011), it is also the main deficit area. Almost all maize production is rain-fed and produced by small farmers, occupying 54 % of small

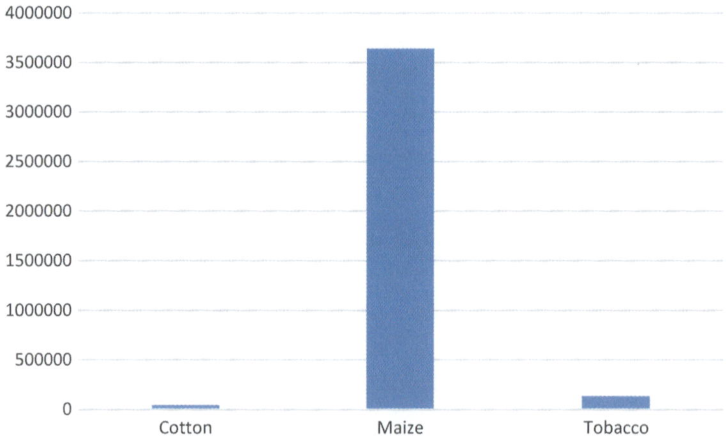

Fig. 4.8 Crops production in Malawi (in tons, 2013). Source: FAOSTAT Online

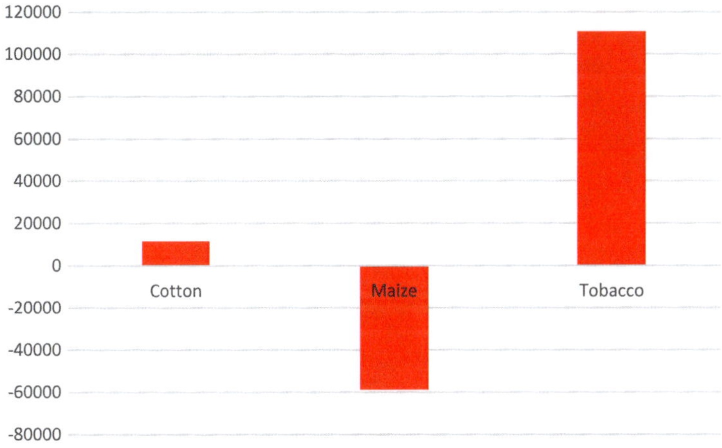

Fig. 4.9 Crops net export in Malawi (in tons, 2013). Source: FAOSTAT Online

producers' cultivated land. The average farm size of smallholder maize producers in Malawi amounts to only 0.5–0.8 ha. The smallest farms are located in the southern region, where population density is higher.

A number of factors can be identified as key constraints to increased maize production levels. Some of the obvious ones are the fact the maize production solely depends on rainfall. This is a major setback to agricultural sector as a whole as a negligible share of agricultural land is irrigated in Malawi. The maize value chain suffers from reduced investment in production as a result of low traded volumes and thin markets. In fact, it is estimated that between 85–90 % of maize is consumed within households and villages, leaving only a small share to be marketed (Dorward and Chirwa 2011). Last, interventions of government in setting minimum price each season is said to cause high price variability for maize sellers, buyers and traders (ibid).

Tobacco is the most important cash crop in Malawi. Tobacco production accounts for over 13 % of Malawi's Gross Domestic Product and provided 62 % of total domestic export earnings in 2010 (National Statistical Office 2010). Burley tobacco is the main variety with a share of over 80 % of total tobacco production. Though traditionally produced by large estates, a series of reforms in the early 1990s increased opportunities for smallholder farmers to participate in the sector (Derlagen 2012). Currently, over 95 % of tobacco is produced by small farmers. Malawi's tobacco production over the last two decades has generally remained steady, from at total production of 130,400 t in 1993 to over 132,800 t in 2013 while hitting an all-time high of 208,000 t in 2009. While tobacco is grown in almost all the regions of Malawi, the central regions of Salima, Lilongwe and Kasungu together accounts for more than half of total production.

One of the key constraints that had been undermining the Malawian tobacco value chain was the market power of leaf merchants who had some form of oligopolistic power over the farmers as explained by Prowse and Moyer-Lee (2013). However, a number of changes were introduced in the value chain during the 2009/2010 marketing season. The first change was the introduction of district markets. In 2009/2010 there were two district markets, one in Kasungu district at Chinkhoma and one in the Southern Region at Ngodi. These markets undermined and effectively replaced the cross-border trade. The second major change was the introduction of minimum prices. This was a result of government effort to ensure that farmers get better prices.

Malawi is not a major cotton producer, but cotton is Malawi's fourth largest agricultural foreign exchange earner behind tobacco, sugar and tea (FAO 2015). The majority of cotton in Malawi is cultivated by approximately 120,000 small holder farmers under rainfed crop cultivation systems (Hardwick 2010). The average landholding per farmer ranges from 0.2 to 1.6 ha and cotton varieties vary according to the region. In 2012, production peaked at 220,726 t, from 52,456 t, attributable to the disbursement of inputs subsidies in non-traditional growing areas (Karonga, Mulanje and Nkhata Bay) in the framework of the FISP (Kenamu and Phiri 2014). However, since farmers who has received fertilizer the previous year were removed the next year, production dropped dramatically in 2013. Most of Malawi's cotton production occurs in regions located in the south. The Lower Shire Valley and Balaka, which represent 50 and 30 % of production, respectively, and the Lakeshore area of Lake Malawi accounts for 20 % (FAO 2015).

Cotton industry faces many challenges in Malawi as illustrated by Gwarazimba (2009). The first challenge is pricing of lint. While stakeholders are engaged in pricing decisions, ginners hardly adhere to agreed prices and as a result production varies annually as farmers change to crops that provide better returns. The second major challenge is low yields. Yields vary between 300 and 700 kgs/ha when variety potential is sometimes in excess of 2000 kgs/ha. The low yields are due to lack of pesticides use and absence of information on cotton husbandry practices. The third challenge is lack of quality seed. Low yields are often attributed to poor quality seed. Another significant challenge is too many varieties on the market. Malawi has three main varieties but there is little control on parent seed availability. Side marketing is also a key challenge to the cotton sector. While farmers are contracted and supplied with inputs, some try to sell their cotton to ginners where they have credit obligations.

Table 4.3 below displays the market shares for processors of cotton, maize, and tobacco in Malawi. Maize is mostly processed informally by a large number of community/village level small scale processors. Of the formal processors, the Agricultural Development and Marketing Corporation (ADMARC) is the largest maize processor in the country. ADMARC is parastatal institutions that has major influence in maize and key commodity prices. Government determine the minimum price for maize. This was reintroduced in 2008 in response to the commodity price shock.

Prior to market liberation, cotton enjoyed the same level of protection offered to cash crops such as tea and tobacco, and was mainly bought buy ADMARC and

Table 4.3 Market shares in Malawi

Maize		Tobacco	
Company	Share (%)	Company	Share (%)
Transglobe	2.00	Alliance One International	34.00
RAB	5.40	Limbe leaf	29.00
ADMARC	8.00	Africa Leaf	16.00
Others	84.60	Premium TAMA	13.00
		Malawi leaf	5.00
		ATC	2.00
		Wallace	1.00
Cotton			
Company	Share (%)		
ADMARC	1.15		
Afrisian	6.27		
Cotton Ginners Africa	17.01		
Cotton Lake Cotton Company	26.46		
Iponga Cotton Company	3.42		
Malawi Cotton Company	30.36		
Nadhi	1.50		
Toleza	11.54		
Woget	2.28		

Source: Stakeholders interviews

processed through a government owned factory. Post liberalization era has seen the influx of private sector and at times formation of cartels which have suppressed farm gate prices. As a result, government introduced a minimum/floor price for the buying of the crop based on the gross margin analysis principles. The largest cotton processor in Malawi, Malawi Cotton Company, controls around 30.36 % of the market followed by Great Lake Cotton Company with 26.46 % and Cotton Ginners Africa with 17.01 %. In the post liberalization era, intermediate buyers were introduced to increase competition in the tobacco marketing. However, some malpractices arose which compromised tobacco quality and inevitably depressing the tobacco prices offered to farmers. As a result, intermediate buyer system was abolished and minimum buying price for tobacco introduced. Today, there are only seven leaf tobacco companies in Malawi. Alliance One International (AIO) is the leading leaf tobacco company in Malawi, controlling 34 % of the market share. Limber leaf controls 29 %, and Africa leaf, the third largest company controls 16 %.

4.3 Simulation Results

In this section, we use our model to perform several simulations consisting of comparative static results stemming from the model. We consider two sets of exercises. Following Porto et al. (2011), we shock the market structure of the supply chain. To this end, we consider (arbitrary) changes in the number of firms and in their market shares to capture variations in the extent of competition in the supply chain. More precisely, we study the case of leaders splitting in half (Leaders Split), the two largest leaders merging (Leaders Merge), Exit of largest firm, Equal market shares and a limit case of Perfect Competition. We also consider comparative static results from changes in key parameters affecting the production decision of the farmers. We explore (arbitrary) changes in international prices, production costs, endowments and risk and food security aversion. We are interested in price changes of the agricultural goods produced in Malawi. We will then use these simulated price changes with our household survey data to assess the welfare and poverty implications.

We investigate the three crops described above in the non-parametric regressions: Maize (importable), Cotton and Tobacco (exportable). Given the complexity of the scenarios, we simplify the analysis by working with a sequence of partial equilibrium models, dealing with each case separately. That is, when we take the case of cotton we leave the maize and tobacco markets unmodeled.

We start with Maize. Maize is an importable food crop by Malawi and therefore we model it as such. Its market concentration is low, the market being fairly competitive. The price changes from the simulation results are presented in Table 4.4. The first row shows the impacts of changes in competition. Given that suppliers are oligopolies, an increase in competition should reduce prices, as it indeed happens. However, the decrease is not large, because market structure consists of many similar small suppliers.

We now focus on the role of household constraints, which are shown in Column 1 of Table 4.4, starting in row 2. This is the baseline model, different rows correspond to different comparative static results. The largest impact come from changes in international prices: if international prices increase 10 %, farmgate prices would increase by 4.69 %. In row 3 we simulate a 10 % increase of the Marginal cost of the cash crop, rival in production to the food crop. The effect is negative, since a less profitable cash crop production makes producers switch to food crops, increasing aggregate supply and therefore pushing prices down. A similar mechanic takes place in row 3, but shocking the Fixed Cost of cash crop instead. In the following row we show the effect on prices of an increase of 10 % in Household Resources, in terms of the model, the endowment. The effect is positive, so prices increase. We would expect a larger endowment to be associated with a stronger supply of various crops. In theory, however, a larger resource base can also facilitate the adoption of the cash crop, which is in principle more profitable, for marginal farmers. If so, maize supply could actually decrease, thus leading to higher prices ceteris paribus. In the 6th row we see the effects of an increase in the Risk and Food Security parameter. With a higher risk aversion, households produce more for autoconsumption than for the market, withdrawing supply and increasing prices.

In the third group of results we study the interactions between different crop markets. We start by shocking the price of the cash crop. Since the cash crop is a rival in production to maize, we would expect households to withdraw supply for

Table 4.4 Farmgate price simulation results for maize

	Baseline	Leader split	Leaders merge	Exit of largest	Equal marquet shares	Perfect Competition
Competition policy	0.00	−0.02	0.02	0.02	0.00	−0.52
Increase of 10 % in						
International price	4.71	4.70	4.72	4.72	4.71	4.48
Marginal cost of producing cash crop	−1.15	−1.12	−1.18	−1.18	−1.15	−0.52
Fixed cost of producing cash crop	−0.12	−0.14	−0.10	−0.10	−0.12	−0.52
Endowment	0.17	0.14	0.21	0.21	0.17	−0.52
Preference parameter	−0.03	−0.05	0.00	0.00	−0.03	−0.52
Cash crop price	1.22	1.14	1.31	1.31	1.22	−0.52
Marginal cost of producing food crop	0.23	0.20	0.27	0.27	0.23	−0.52
Non-farmer demand	0.01	−0.01	0.04	0.04	0.01	−0.52

Source: Simulation results from the model of Chap. 2

maize and increase supply of cash crop, leading to higher prices for maize. This is indeed what we show in row 7. An increase in marginal cost of food crop (maize) has the opposite effect of an increase of the marginal cost of cash crop mentioned above; households switch from maize to the cash crop as the former becomes unprofitable and prices increase. Finally, in the last row we show the effect of an increase of food demand, which unsurprisingly increases the price of food crop.

We note that the effects seem to be very small, with the exception of competition policy and the increase in international prices. This is to be expected. The underlying logic of the model is that of a market of a tradable good, in which the country is a price taker. In such a model, internal demand/supply shifts affect only quantities: production, consumption, imports and exports, but never prices because the price always equals the international price, which is given. The main difference between our model and the more basic model is that we simulate suppliers, which create a gap between international and local prices. This gap is strongly affected by market concentration, but little by anything else: this is why we see the highest variations when international prices increase (meaning the gap remains mostly constant) but not in internal shocks to parameters, which are the equivalent of internal demand/supply shifts. This implies that quantities and not prices do most of the adjustment.

The fact that quantities are the adjustment variable for these parameter shocks also implies that the results we show below could be greatly underestimating the true welfare effects. Remember that the results we show are first-order approximations to the true welfare effects, which ignore the increase in welfare measured by the triangle resulting from the increase of quantities times the new price minus the increase in costs associated to that increase in quantities. Therefore if quantities and not prices is the adjustment variable, what we omit is likely to be more relevant than what we include. Prices do remain the best approximation, but a very inaccurate one.

We can also analyze complementarities and substitutabilities. However, the possibilities of generalizing results are limited. We say two policies are complimentary when the joint effect of the two policies is larger than the sum of the separate effects. Consider for example Leaders Merge with the increase of cash crop prices. The sum of the separate effects is 0.93, whereas the joint effect is larger, 0.98. On the other hand, two policies are substitutable when the joint effect of the two policies is smaller than the sum of the separate effects. For example perfect competition and an increase in the marginal cost of the cash crop. The separate effects sum −1.04 but the joint effect is half of that −0.52. In this case the effect is solely that of perfect competition, since if market structure converges to perfect competition, the model converges to that of a tradable good, and the internal price equals the international prices. As we explained above, under such model all internal shocks affect only quantities, and not prices, making the increase of local marginal costs completely irrelevant for our analysis.

It is noteworthy that no clear pattern emerges with regards to substitutabilities and complementarities. Sometimes two shocks oppose each other, or go in the same direction. Sometimes the joint effect is much larger than the separate effect, sometimes it is not. Irrespective of the actual results, what we intend to stress with these exercises is that complementarities and substitutabilities do exist, they can be derived by a model, and should be taken into account when designing agricultural policies.

We turn now to the case of Cotton and Tobacco. Unlike Maize, these are export cash crops. We report the main results in Table 4.5 for Cotton and Table 4.6 for Tobacco. Another difference is that their market structure is much more concentrated than in Maize. The two largest firms amount to 30 and 26 % of total market for Cotton and 34 and 29 % for Tobacco. The elimination of such market power creates not only much larger price variations, but also internal shocks with stronger effects, for the gap between international prices and local prices is larger, and therefore the quantitative effect of the adjustment of this larger gap by oligopsonies has a larger effect on farmgate prices.

We summarize the results. In row one we see that if the Leader would split, farmgate prices would rise 1.37 % and 1.11 % for Cotton and Tobacco respectively. Prices increase and not decrease like with Maize, for this is an oligopsony and not an oligopoly, that is firms are marking the price down and not up. Consistently, if Leaders Merge Cotton and Tobacco farmgate prices fall 1.46 and 1.02 % respectively. Farmgate prices for the two crops fall again in the case of Exit of the largest firm, 1.71 % for Cotton and 1.18 % for Tobacco. Farmgate prices increase 5.66 % for Cotton and 4.41 % for Tobacco if the two largest firms were to have Equal Market shares. Finally, were the market structure to converge to Perfect competition, farmgate prices would rise 12.71 % for Cotton and 8.94 % for Tobacco.

Table 4.5 Farmgate price simulation results for cotton

	Baseline	Leader split	Leaders merge	Exit of largest	Equal market shares	Perfect competition
Competition policy	0.00	1.37	−1.46	−1.71	5.66	12.71
Increase of 10 % in						
International price	17.56	19.29	15.52	15.27	23.66	32.71
Marginal cost of producing cash crop	1.09	2.34	−0.19	−0.42	6.50	12.71
Fixed cost of producing cash crop	0.31	1.66	−1.10	−1.34	5.96	12.71
Endowment	−0.94	0.50	−2.49	−2.71	4.66	12.71
Preference parameter	1.21	2.68	−0.14	−0.40	7.05	12.71
Food crop price	0.95	2.08	−0.25	−0.48	6.36	12.71
Marginal cost of producing food crop	−0.31	1.08	−1.88	−2.13	5.44	12.71
Non-farmer demand	0.00	1.37	−1.46	−1.71	5.66	12.71

Source: Simulation results from the model of Chap. 2

Table 4.6 Farmgate price simulation results for tobacco

	Baseline	Leader split	Leaders merge	Exit of largest	Equal marquet shares	Perfect competition
Competition policy	0.00	1.11	−1.02	−1.18	4.41	8.94
Increase of 10 % in						
International price	13.09	14.34	11.70	11.50	17.83	23.86
Marginal cost of producing cash crop	2.06	3.05	1.03	0.82	5.65	8.94
Fixed cost of producing cash crop	0.20	1.34	−0.76	−0.94	4.58	8.94
Endowment	−0.58	0.43	−1.65	−1.84	3.77	8.94
Preference parameter	1.30	2.51	0.13	−0.05	5.57	8.94
Food crop price	1.19	2.24	0.24	0.08	4.96	8.94
Marginal cost of producing food crop	−0.14	0.96	−1.25	−1.42	4.28	8.94
Non-farmer demand	0.00	1.11	−1.02	−1.18	4.41	8.94

Source: Simulation results from the model of Chap. 2

Overall, the price impacts tend to be larger for Cotton than for Tobacco, for its market is more concentrated.

If we look at row number two, we see that for Cotton, a 10 % increase in international prices leads to a 17.56 % increase in farmgate prices, whereas for Tobacco the increase in farmgate prices is 13.09 %. Here results do not change with respect to those in Maize: an increase in international prices will always lead to an increase in farmgate prices, irrespective of market structure or whether it's an importable/exportable food/cash crop.

Results do change when we take into account internal shocks. In row 3 we show that increases in the marginal cost of the cash crop lead to increases in the farm gate price of both Cotton and Tobacco of 1.09 and 2.02 %, since we are talking now of the increase of the cost of the good being produced and not the rival good. Like in Maize, an increase in fixed costs of 10 % also increases the price of the crops, 0.30 % for Cotton and 0.20 % for Tobacco (row 4). In row 5 we show that an increase in household resources lowers prices as more farms enter production, like with Maize, the decreases being 0.94 and 0.58 % for Cotton and Tobacco respectively. An increase in risk aversion pushes prices up as households switch to producing food for autoconsumption and internal supply of the cash crop falls. Row 6 shows that these price increases correspond to 1.21 % for Cotton and 1.30 for Tobacco.

Finally we focus now on the shocks to the alternative food crop, in rows 7–9. A rise of the food crop price makes the farmgate price increase: 0.95 % in the case of Cotton and 1.19 % as row 7 shows. If the marginal cost of producing the food crop increases 10 %, now the price of the cash crop under analysis falls rather than increases unlike with maize, because maize was the food crop before, but now we are dealing with cash crops. In the current case, the food crop is a rival in production, then if the food crop's cost increases, firms switch from food to cash crop production, pushing the cash crop's price down. Row 8 shows that this fall is 0.31 and 0.14 % for Cotton and Tobacco respectively. Finally, we show that an increase in non-farmer demand has no effect on the price of the cash crop. This is because we are compounding two very weak effects. Firstly, an increase in food crop demand leads to an increase in the price of the food crop, which is quantitatively negligible as we have seen in row 9 of Table 4.4 for Maize. Secondly, that price increase affects the price of the cash crop, as in row 7 of Table 4.5 for Cotton and Table 4.6 for Tobacco. It is then unsurprisingly that the effect is zero, although had the effects been larger it would have been positive.

4.4 Welfare Simulations

We complete our analysis by discussing the welfare and poverty impacts of the comparative static results presented above. The ultimate focus of our research is the role of supply chains in agriculture on household welfare, both on average and for specific groups such as the poor and non-poor. We are also interested in whether it is possible to inform policy about ways to ameliorate the poverty impacts by analyzing complementarities between the structure of markets and household constraints. This is the goal of this section.

To perform our analysis we use standard techniques in the literature. Namely, the first-order approximation of Deaton (1989a, b, 1997). As we discussed in the previous section, the first order approximation works well if the price changes are

Table 4.7 Maize price changes and household welfare

	Baseline	Leader split	Leaders merge	Exit of largest	Equal market shares	Perfect competition
Total						
Competition policy	0.00	0.00	0.00	0.00	0.00	0.03
International price	−0.27	−0.27	−0.27	−0.27	−0.27	−0.26
Poor						
Competition policy	0.00	0.00	0.00	0.00	0.00	0.05
International price	−0.42	−0.42	−0.42	−0.42	−0.42	−0.40
Non poor						
Competition policy	0.00	0.00	0.00	0.00	0.00	0.02
International price	−0.17	−0.17	−0.17	−0.17	−0.17	−0.16
Producers						
Competition policy	0.00	0.00	0.00	0.00	0.00	−0.01
International price	0.06	0.06	0.06	0.06	0.06	0.06

Note: First order impact on household welfare

small and if there are limited supply and consumptions responses. It is, in general, a very powerful and useful tool to evaluate the welfare effect of price changes.

The welfare impacts of price changes are shown in Tables 4.7, 4.8 and 4.9 for Maize, Cotton and Tobacco. We show the impacts of shocks to the market structure. To illustrate the complementarities, we show results for a combination of shocks to market structure and international prices, commenting on the results for other complementarities at the end. We report average as well as disaggregated results for specific groups: poor/non-poor and producers.

We start with row 1, which shows the overall effect of competition policy. We see that in all three crops more competition would increase welfare. An interesting results emerges from the analysis, comparing the welfare impacts with the comparative static. Although cotton had the highest price change because of competition policy, it is tobacco's competition policy that would increase the welfare of household the most, because it is a more relevant (defined as: with larger net budget shares) than cotton.

In row 2 we analyze the overall effect of price increases. Since Maize is an imported crop, the average net position of households is buyer, therefore a price increase reduces overall welfare. The opposite is true for cotton and tobacco. It is also noticeable how maize's price change has a higher effect (in absolute value) than cotton, because of its relevance in the household's budgets.

Table 4.8 Cotton price changes and household welfare

	Baseline	Leader split	Leaders merge	Exit of largest	Equal market shares	Perfect competition
Total						
Competition policy	0.00	0.00	−0.01	−0.01	0.02	0.04
International price	0.06	0.07	0.05	0.05	0.08	0.11
Poor						
Competition policy	0.00	0.01	−0.01	−0.01	0.02	0.05
International price	0.07	0.08	0.06	0.06	0.10	0.13
Non poor						
Competition policy	0.00	0.00	0.00	−0.01	0.02	0.04
International price	0.05	0.06	0.05	0.05	0.07	0.10
Producers						
Competition policy	0.00	0.16	−0.17	−0.19	0.64	1.45
International price	2.00	2.20	1.77	1.74	2.70	3.73

Note: First order impact on household welfare

Row 3 and 4 shows the effect of competition policy and international price changes on the poor, and rows 5 and 6 for the non-poor. Interestingly, each of the possible combination is reflected on the crops under analysis. For Maize, the effect on the poor is roughly three times the effect on the non-poor, although both effects are very small. Policies to foster competition in the maize market would then be pro-poor. On the contrary, price increases in the maize markets would have a negative average effect, and more strongly negative for poorer households. For Cotton, the effect of competition policies and price increase is equal among the poor and the non-poor, meaning policy directed at Cotton will have no effects on income inequality. Tobacco is the other possible case: policy affects the non-poor twice as much as it affects the poor. Therefore, policy directed at increasing competition will tend to make the income distribution more unequal, as would price increases.

Finally, rows 7 and 8 show the effects of competition policy and price increases on the welfare of producers. This allows us to have a grasp of how diversified households are in the consumption/production of the crops, if the average calculated first is not averaging away sharp variations of welfare for some households. This does not seem to be the case for Maize, producers are actually hurt the least, less than poor and non-poor. This might be because maize producers produce mostly for auto consumption, therefore price variations do not affect them much. This is not the case for both Cotton and Tobacco. Tobacco producers are less

Table 4.9 Tobacco price changes and household welfare

	Baseline	Leader split	Leaders merge	Exit of largest	Equal market shares	Perfect competition
Total						
Competition policy	0.00	0.04	−0.04	−0.05	0.17	0.36
International price	0.52	0.57	0.46	0.46	0.71	0.95
Poor						
Competition policy	0.00	0.03	−0.03	−0.03	0.11	0.22
International price	0.32	0.35	0.29	0.28	0.44	0.59
Non poor						
Competition policy	0.00	0.06	−0.05	−0.06	0.22	0.45
International price	0.66	0.72	0.59	0.58	0.89	1.20
Producers						
Competition policy	0.00	0.30	−0.27	−0.32	1.18	2.40
International price	3.51	3.85	3.14	3.09	4.79	6.41

Note: First order impact on household welfare

diversified than cotton producers facing larger welfare effects with lower actual price changes.

These results are in line with the literature on the topic (see the review in Lederman and Porto 2015). There are several elements to be taken into account. Firstly, first-order approximation through price-changes in shocks that mainly affect quantities, which we discussed in Sect. 3.2. Secondly, in these welfare changes we are taking only the net position of households. Some may be relevant on the consumption side, and on the production side. But we are only interested in the net position of the household, netting those effects out. Thirdly, for maize the market structure was almost competitive, leaving no room for sizable price changes. The combination of small price changes with small net differences implies small impacts (Deaton 1997).

The fact that the impacts calculated are small does not mean that the welfare impacts are not important. As we argued above, small results are expected in this literature. They are expected given the context (household survey and baseline market structure) but are reasonable. We are only taking into account short run impacts of price changes cause by variations of exporter's market power with complementary factors. We note that these factors have an independent effect on welfare that we do not measure. For example, if the cost of crop production declines because of infrastructure improvement, there will be a direct impact on welfare and an indirect one through the combination with changes in market structure. Here we measure the direct impact only. It turns out that these additional impacts are small but, since they do not carry additional costs (such as fiscal costs if the complementarities are government-funded) they only generate benefits.

References

Deaton A (1989a) Rice prices and income distribution in Thailand: a non-parametric analysis. Econ J 99(395):1–37

Deaton A (1989b) Household survey data and pricing policies in developing countries. World Bank Econ Rev 3:183–210

Deaton A (1997) The analysis of household surveys: a microeconometric approach to development policy. Johns Hopkins University Press for the World Bank

Derlagen C (2012) Analysis of incentives and disincentives for maize in Malawi. Technical notes series, MAFAP, FAO, Rome

Dorward A, Chirwa E (2011) The Malawi agricultural input subsidiy programme: 2005–6 to 2008–9. Int J Agric Sustain

FEWSNET (2011) Southern Africa cross border trade update, April 2011

Food and Agriculture Organization (FAO) (2015) Malawi country profile. FAO, Rome

Gwarazimba V (2009) Cotton and cassava seed systems in Malawi, Mozambique, and Zambia. In: Paper commissioned by FAO under the All ACP Agricultural Commodities Programme

Hardwick T (2010) Quantitative value chain analysis, an application to Malawi. Policy research paper, World Bank

Kenamu E, Phiri A (2014) Performance of cotton production in Malawi. Scholarly J Agric Sci

Kharas H, McArthur J, Gertz G, Mowlds S, Noe L (2015) Ending rural hunger: mapping needs and actions for food and nutrition security. Global economy and development. Brookings Institution, Washington, DC

Lederman D, Porto G (2015) The price is not always right: on the impacts of (commodity) prices on households (and countries). World Bank Res Obs 31(1):168–197

Mcmillan M, Rodrik D (2011) "Globalization, structural change and productivity growth", 2011. In: Bachetta M, Jansen M (eds) Making globalization socially sustainable. International Labor Organization, Geneva

National Statistical Office (2010) Statistical year book—2010. Zomba, Malawi

Porto G, Depetris Chauvin N, Olarreaga M (2011) Supply chains in export agriculture, competition, and poverty in Sub-Saharan Africa. CEPR Press, London

Prowse M, Moyer-Lee J (2013) A comparative value chain analysis of smallholder burley tobacco production in Malawi—2003/4 and 2009/10. J Agrar Chang 14(3):323–346

Chapter 5
The Case of Zambia

This chapter analyzes the case of Zambia. We describe the household survey (Sect. 5.1), the food and cash crops institutional arrangements (Sect. 5.2) and we present the farmgate price simulation results from the model (Sect. 5.3) and the household welfare simulations (Sect. 5.4).

5.1 The Household Survey Data

Zambia's data stems from the Living Conditions Monitoring Survey of 2003 which covers more than 9000 households. Observations are evenly split among the rural and urban categories, 48.9 % being rural and 51.1 % being urban, whereas the actual population is majorly rural (65.1 %). The population is very young: 45.7 % are 14 years old or less, and only 2.8 % are 65 or older. When it comes to gender the population is balanced: 50.9 % are female and 49.1 % are male. This balance among the sexes is broken when it comes to household heads: only 23.1 % of household are headed by females, 24.1 % in rural areas and 21.3 % in urban ones. Household sizes are big: 6.77 people per household on average, 3.8 being under 18 years old. Household sizes do not differ significantly between rural and urban areas (6.77 and 6.68 respectively).

Having described demographic data, we now focus on economic variables. Figure 5.1 shows the distribution of income. The graph shows the estimated density function of the logarithm of household per capita expenditure at the national level and for urban and rural regions separately. As usual, the distribution for urban households lies to the right of the rural distribution, meaning urban households have a higher expenditure level than rural ones on average. Since there are more rural households than urban ones, the national distribution is closer to the rural than to the urban distribution.

N. Depetris Chauvin et al., *Agricultural Supply Chains, Growth and Poverty in Sub-Saharan Africa*, Advances in African Economic, Social and Political Development, DOI 10.1007/978-3-662-53858-6_5

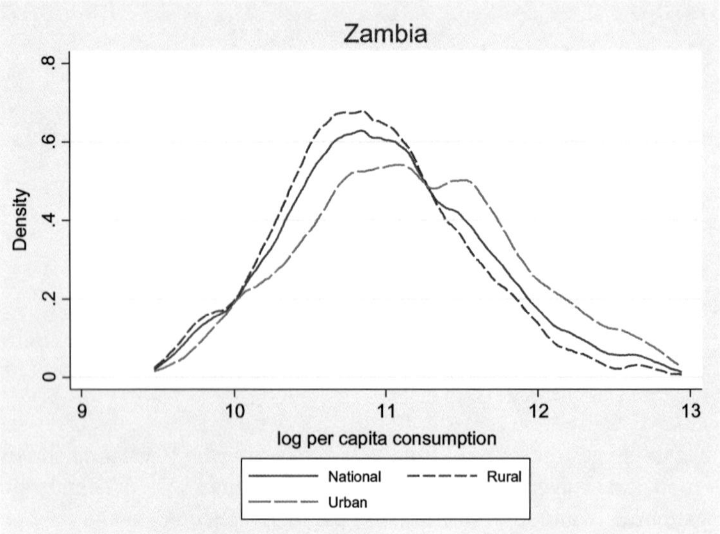

Fig. 5.1 The distribution of income density of (log) per capita household expenditure. Source: Zambia Living Conditions Monitoring Survey (2003)

We now describe the sources of income and the patterns of consumption across households. In order to do this, we turn to Table 5.1, where we report consumption patterns both for urban and rural regions, including cash expenditures and the value of autoconsumption. The latter is much higher than the former in rural areas compared to urban areas, which is to be expected. Cash expenditures are higher than autoconsumption even for rural households. We focus on the share of food consumption in the households' budgets, which is similar in rural and urban households, 65.6 % and 57.5 % respectively. This is coherent with our previous data on the distribution of expenditures presented above and Engel's law: urban households are richer than rural households, and they therefore spend a smaller share of their budget in food. The most relevant crop we have in the data is maize, amounting to 15.8 % of total expenditures, 18.1 % for rural and 11.5 % for urban household. Cash crops are insignificant in consumption, as expected.

In Table 5.2, we focus on the income side. As it is usually the case, autoconsumption is much larger in rural households than in urban households (60.5 versus 9 %). Disaggregating the data reveals that maize is the most significant crop, with 16.9 % of total income for rural and 2.3 % of total income for urban households. The second crop in importance is poultry, consisting of 6 % of income on average for rural households and only 0.4 % for urban households.

To analyze the welfare effects of changes in the prices of these commodities, we must first describe the patterns of income and expenditure sources along the income distribution. We will take the log of per capita household expenditure (log pce) as a measure of income and we plot estimates of non-parametric regression of income and budget shares on log pce.

Table 5.1 Budget shares

Zambia	Total	Rural	Urban
Total consumption per capita	100.0	100.0	100.0
Expenditures	69.0	54.6	96.3
Food	31.7	20.2	53.7
Manufactures	0.0	0.0	0.0
Services	0.0	0.0	0.0
Others	37.2	34.4	42.5
Auto-consumption	31.0	45.4	3.7
Auto-consumption food	31.0	45.4	3.7
Auto-consumption others	0.0	0.0	0.0
Total Food consumption	62.8	65.6	57.5
Total crops	28.4	32.5	20.7
Maize	15.8	18.1	11.5
Rice	1.5	1.4	1.5
Poultry	3.6	4.9	1.0
Livestock	5.2	6.0	3.5
Wheat	2.1	1.6	3.0
Tobacco	0.3	0.4	0.1
Cotton	0.0	0.0	0.0

Source: Zambia Living Conditions Monitoring Survey (2003)

Table 5.2 Income shares

Zambia	Total	Rural	Urban
Total income per capita	100.0	100.0	100.0
Incomes	56.9	39.5	91.0
Food (agriculture)	6.3	3.6	1.9
Wage	21.1	6.6	49.3
Enterprises	14.7	9.3	25.0
Transfers	14.9	15.0	14.8
Auto-consumption	43.1	60.5	9.0
Auto-consumption food	43.1	60.5	9.0
Auto-consumption others	0.0	0.0	0.0
Total food income and AC	49.4	69.0	10.9
Total crops	24.6	35.0	4.3
Maize	12.0	16.9	2.3
Rice	0.4	0.5	0.0
Poultry	4.1	6.0	0.4
Livestock	2.6	3.8	0.2
Wheat	0.1	0.1	0.0
Cassava	0.5	0.7	0.1
Millet	0.1	0.1	0.0
Yam	3.5	4.7	1.3
Tobacco	0.3	0.5	0.0
Cotton	1.0	1.6	0.0

Source: Zambia Living Conditions Monitoring Survey (2003)

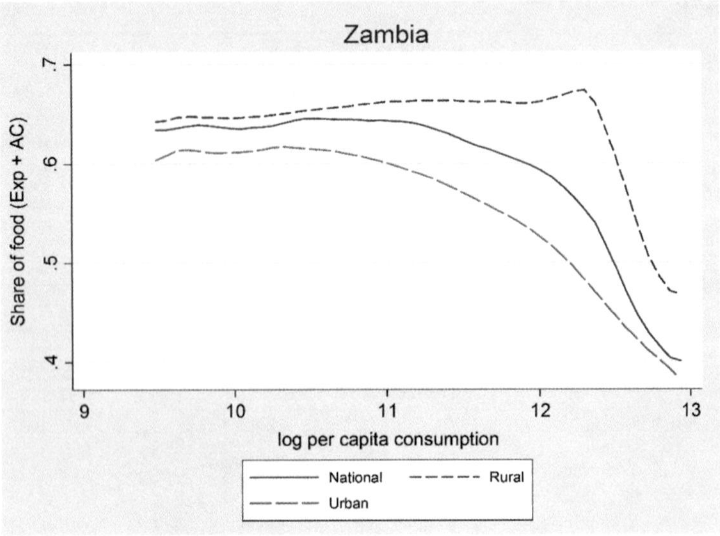

Fig. 5.2 Total food budget share across the income distribution. Source: Zambia Living Conditions Monitoring Survey (2003)

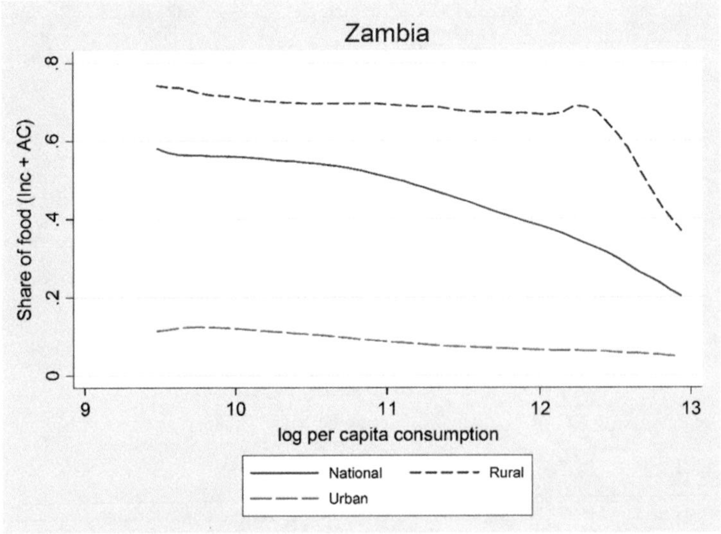

Fig. 5.3 Total food income share across the income distribution. Source: Zambia Living Conditions Monitoring Survey (2003)

Figures 5.2 and 5.3 refer to the average share of food expenditure and the average share of food income, including cash expenditures as well as auto-consumption. In both cases, food share slopes downward, staying mostly stable to fall abruptly as income increases. Since we see the same abrupt fall both in income and expenditures, this implies that this fall is caused by a shift out of autoconsumption.

For food expenditures, the abrupt fall is led by rural households, who switch out of food consumption abruptly after a certain threshold, whereas urban households move out more smoothly as their income increases. However, even at the top of the income distribution food expenditure remains high: 50 % for rural households and 40 % for urban ones.

For food income the pattern is different. Urban households derive little of their income from food production compared to rural ones. Urban households' food share curve slopes downward slowly and smoothly, whereas rural households' curve does so up to a threshold where it falls sharply. Rural households at the top of the income distribution derive 40 % of their income from food production, but this number is low compared to those rural households at the bottom of the income distribution: 80 %. For urban households income derived from food production is not insignificant as well: 20 % for those at the bottom of the income distribution and less than 10 % for those at the top.

We now disaggregate our analysis by studying specific crops. We see that for Maize, both in the expenditure (Fig. 5.4) and in the income side (Fig. 5.5), rural households have much larger shares than urban ones. Moreover, on the income side, maize shares seem to be increasing in income, while the opposite is true for urban households. The abrupt fall of shares in food income described above seems to be caused by a sharp fall in maize shares for the wealthiest individuals.

We present non-parametric regressions of cash crops as well, ignoring the expenditure side as the shares are close to zero, if not zero. The income side is

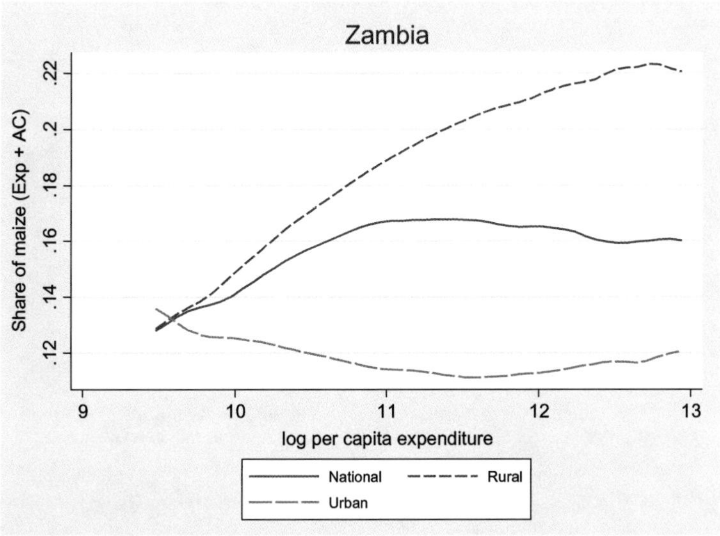

Fig. 5.4 Maize budget share across the income distribution. Source: Zambia Living Conditions Monitoring Survey (2003)

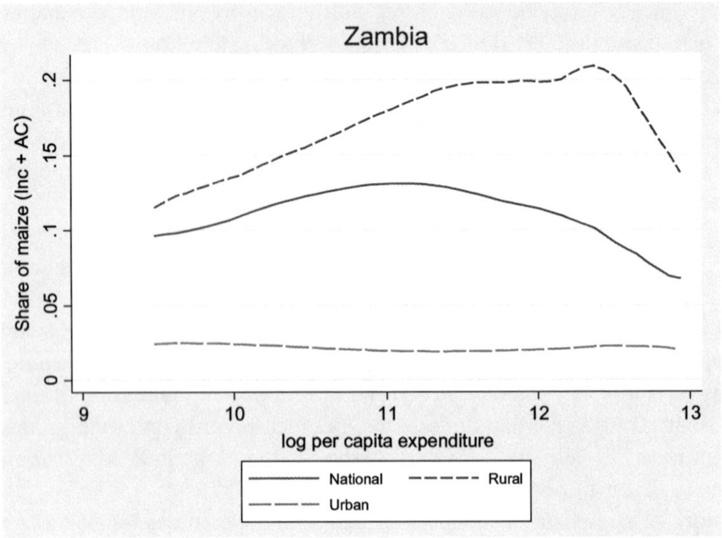

Fig. 5.5 Maize income share across the income distribution. Source: Zambia Living Conditions Monitoring Survey (2003)

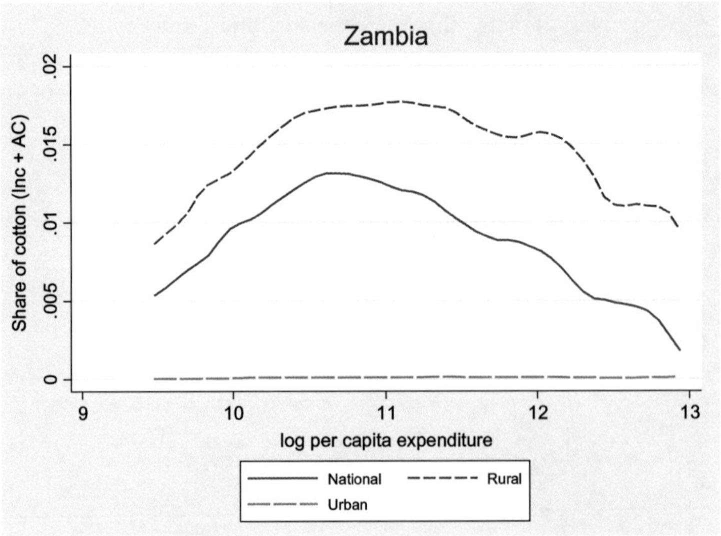

Fig. 5.6 Cotton income share across the income distribution. Source: Zambia Living Conditions Monitoring Survey (2003)

shown in Figs. 5.6 and 5.7. Figure 5.6 corresponds to the cash crop Cotton. The shares follow an inverted u pattern, the highest shares corresponding to the middle of the income distribution. As expected, urban households do not produce the crop. Figure 5.7 shows the income patterns for Tobacco. As in other countries like

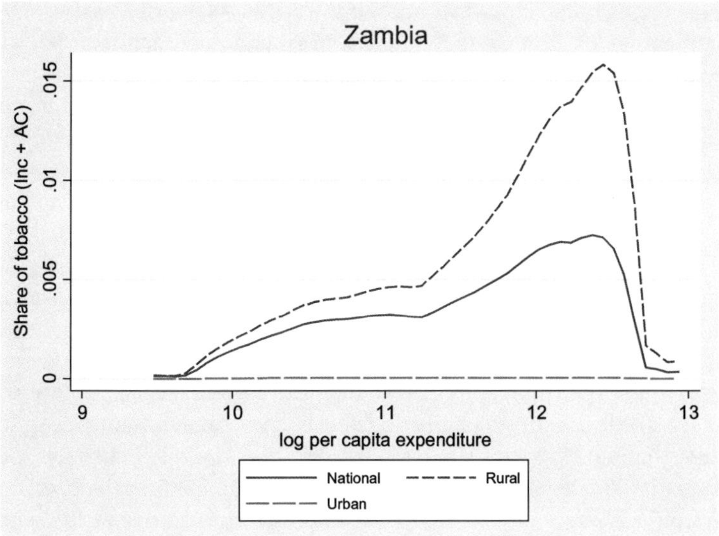

Fig. 5.7 Tobacco income share across the income distribution. Source: Zambia Living Conditions Monitoring Survey (2003)

Malawi, tobacco seems to be strongly pro-rich, its shares increasing steadily along the income distribution, but to fall abruptly at the very top.

5.2 Food and Cash Crops in Zambia

Agriculture in Zambia supports the livelihoods of over 70 % of the population. Zambia's economy has grown steadily in real terms since 2001. However, the percent contribution of the agricultural sector to GDP has declined from 16 % in 2001 to 9.6 % in 2013. Despite a relatively low population density, growth in the number of rural households contributes to increasing land fragmentation and shrinking land size holding in Zambia. It has an agricultural land of 237,360 km^2, of which agricultural land as a percentage of the land area stands at 31.9 % (World Development Indicators 2016). While the mean land size holding in Zambia is 3.27 ha, a quarter of the rural population controls on average barely one hectare of land. Small-scale farming systems in Zambia are dominated by maize production. In 2009/2010, 81.72 % of all smallholders grew maize. Cassava cultivation, the second most important staple food crop, is geographically confined to the north and northwestern parts of Zambia. Groundnuts, the second most widely cultivated crop in Zambia and important source of protein in Zambian diets, is frequently intercropped with maize. In Zambia, groundnuts are often considered a "women's crop" due to their importance for home consumption. Other crop cultivated includes vegetables and wheat. Cotton and Tobacco are two of the most important cash crops in Zambia. Yields for all crops in Zambia are well below global averages. However, while national yields are low, the top 10 % of smallholders achieve yields that are one to nearly four metric tons more than the average depending on the crop. This suggests the potential for yield improvements in Zambia. While input use has trended upward since 2001, 60 % of Zambia farmers still do not use fertilizer on their fields, while more than 60 % do not use hybrid maize seeds.

The current major guiding policy document of the agriculture sector in Zambia is the Sixth National Development Plan (SNDP) 2011–2015, aimed toward attainment of sustained economic growth and poverty reduction. The SNDP identifies agriculture, livestock and fisheries as key priority growth sectors along with mining, tourism, manufacturing, commerce and trade. Another policy driver is the National Agricultural Policy (2004–2015) of Zambia aimed at ensuring that food security is achieved through year-round production and post-harvest management of adequate supplies of basic foodstuffs at competitive costs. The Zambia National Agricultural Investment Plan 2014–2018, which is directly aligned to the national agricultural policy supports the development of sustainable, dynamic, diversified and competitive agricultural sector. It also aims to support food security at household and national levels and maximizes the sector's contribution to GDP. There have been offshoot interventions from these policies. One prominent one is the Farmer Input Support Program (FISP) introduced in 2008. This program supports resource constraint small holder farmers in accessing subsidized inputs, mainly seeds and fertilizer through an e-voucher system. The impact of the program has been mixed, for instance Mason and Tembo (2015) report that although FISP fertilizer raises incomes and reduces the severity of poverty, the program has no statistically significant effect on poverty incidence among smallholder farm households in Zambia. Specifically, a 200-kg increase in FISP fertilizer raises total household income by approximately 7.7 % and reduces poverty severity by 3.6 % points, but these effects are not enough to reduce the probability of household income falling below the poverty line.

We have selected to analyze three crops for the case of Zambia. These include, maize, cotton, and tobacco. We want to see how price changes are sensitive to crop type and we have chosen these three crops because they represent the cash/food crop dichotomy. Maize is the most important grain crop in Zambia, being both the major feed grain and the staple food for the majority of the population. Zambia achieved a historic harvest for the 2013/2014 farming season when over 3.35 million metric tonnes of maize were harvested. This strong performance has been attributed to an 8.1 % increase in the area planted and 22.3 % increase in yields (KPMG 2015). Additionally, the sector has benefitted substantially from the FISP, which is normally targeted at maize growers and a huge share of the agricultural budgetary allocation, about 60 %. Maize is grown in most parts of the country, predominantly in the Eastern, Central and Southern provinces of the country. Smallholder maize farmers account for 80 % of all production. Nearly 50 % of rural households are net buyers of maize in Zambia (Kuteya 2012).

Maize production faces a number of constraints. Aside from the fact that production of maize is rain-fed, there is market unpredictability between the formal and informal maize markets. Competition in maize market requires the development of a predictable set of rules and regulations regarding government's intervention in the maize market. Kuteya and Sitko (2015) recommends improving predictability by weakening the capacity of political actors to direct the Food Reserve Agency (FRA), which is involved in the marketing of maize in Zambia. They propose locating the FRA within the Central Bank or creating a council drawn from the private and public sector to guide its actions (Kuteya and Sitko 2015).

Cotton is grown in Southern, Central, Lusaka, Eastern, parts of the Copperbelt and the Western Provinces. Cotton production in Zambia is mostly rain-fed. Zambia is mainly a raw cotton exporter. It is mainly produced as a cash crop by more than 300,000 small scale farmers under out-grower contracts with the ginning companies. Scope exists for commercial farming as vast tracts of land are available in areas where cotton production is conducive. Cotton production declined marginally in the 2013/2014 to 100,106 MT compared to the 102,938 MT produced in the 2012/2013. Zambia's cotton lint exports plummeted by 57 % in the 2012/2013 to 39,524 MT after 92,723 MT of cotton lint where exported in the 2011/2012. Cotton lint consumption by Zambia's spinning industry is approximately 14,500 MT, while local lint production of 40,000 MT exceeds local demand. South Africa remains the most import destination for Zambian cotton after a major decrease in cotton exports to China.

Cotton production is not without its share of constraints. Productivity of the cotton sector is low, with an estimated national average of between 550–800 kg/ha of seed cotton. Also the inconsistent Government policy on provision of agricultural inputs like the FISP and the FRA support for maize production swings the pendulum against smallholder cotton farmers. Also, out grower schemes have been affected by side selling by cotton farmers due to various ginning companies who have entered the market. Furthermore, increased production in major world markets due to subsidies and use of bio-technology in cotton production undermine the competitiveness of Zambia's cotton in international markets.

Tobacco production is a very attractive investment opportunity in Zambia because it is 7.5 times more profitable per hectare than maize production and 14 times more profitable than cotton (Kuteya 2012). Tobacco production is comprised mainly of out-grower schemes. The crop farm sizes range up to 1000 ha, but the majority grow crops under 5 ha or between 80 and 120 ha. The Eastern, Central and Southern Provinces of Zambia provide the best land and climatic conditions for growing tobacco.

The production of tobacco faces a number of constraints. Prominent among them is the use of low technology by smallholder farmers. Production is rain-fed, and this affects yields. There is also delays in payments of farmers by the licensed tobacco buying companies, which affects production. Further, a lack of clear consistent policy direction has crowded out some licensed tobacco companies who have relocated to neighboring countries like Malawi and Zimbabwe to ply their trade (Figs. 5.8 and 5.9).

The table below displays the market shares for processors of maize, cotton and tobacco in Zambia. The Zambian Government runs a mostly liberalized economy; it has no direct policy to fix prices of agricultural commodities but it may indirectly affected them by specific interventions. Maize is traded mostly informally. Of the

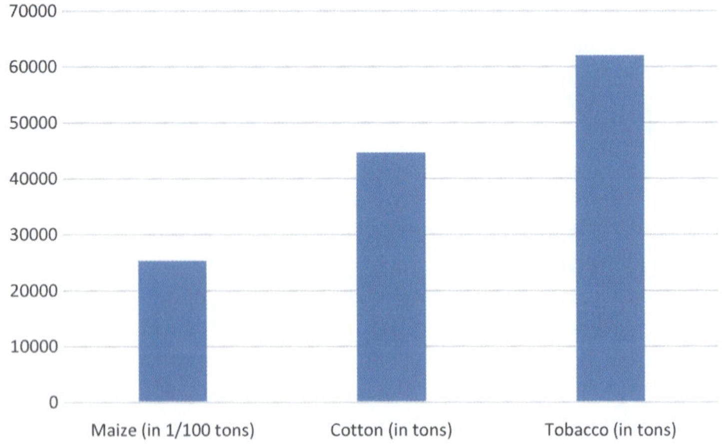

Fig. 5.8 Crops production in Zambia (in tons, 2013). Source: FAOSTAT online

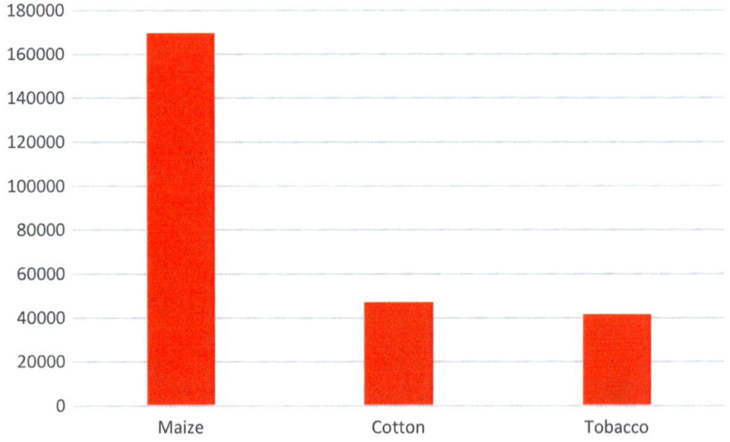

Fig. 5.9 Crops net export in Zambia (in tons, 2013). Source: FAOSTAT online

formal millers, the national milling company has a market share of 22 %, while Antelope milling company controls 20 % of the formal market share. But there is the involvement of the Food Reserve Agency in the marketing of maize in Zambia. They purchase maize from farmers in deprived areas of the country, and they usually announce the price at which it will buy maize from these local farmers. The government does this by offering higher prices than those offered by the private traders in order to ensure higher returns for these farmers. The government also regulates the export and import of maize. Export volumes are determined once domestic maize production has been measured for the year. Similarly, the price of cotton is determined based on market forces without government interference. However, there is the presence of the Cotton Association of Zambia (CAZ), which was created by the Zambia National Farmers' Union (ZNFU) to represent farmers'

Table 5.3 Market shares in Zambia

Maize		Tobacco	
Company	Share (%)	Company	Share (%)
National Milling Company (NMC)	22	Zambia Tobacco Leaf	47.61
Antelope Milling	20	Alliance One Tobacco	16.34
APG Milling	14	Tombwe Tobacco	25.21
Choma Milling	5.60	Associated Tobacco	10.84
Others	38		
Cotton			
Company	Share (%)		
Dunavant	47.23		
Cargill	24.64		
Great Lakes	4.11		
Alliance Cotton	3.29		
Continental	10.27		
Mulungushi	4.11		
Chipata-China Cotton	6.16		
Mukuba	0.21		

Source: Stakeholders interviews

interests in the cotton sector as well as providing the Cotton Ginners' Association with an organized private sector body to communicate on key issues affecting them and the sector. Dunavant, the largest cotton processor in Zambia control 47.23 % of the market, while Cargill controls 24.64 %. Prices for different grades of tobacco are set by the tobacco companies at the beginning of the buying season. However, when the price offered by these traders is not favorable for the farmers, they enter into confrontation with the Tobacco Board of Zambia (TBZ) so that the buyers can review their prices. The TBZ does not control tobacco prices in the country but regulates the rest of tobacco activities as prices are determined by market forces at international level. There is however a price floor that is set based on the international traded price of tobacco. The Zambia Tobacco leaf company controls 47.61 % of the market, while the Alliance One Tobacco company controls 16.34 % (Table 5.3).

5.3 Simulation Results

In this section, we use our model to perform several simulations consisting of comparative static result derived from the model. We consider two sets of exercises. Following Porto et al. (2011), we shock the market structure of the supply chain. To this end, we consider (arbitrary) changes in the number of firms and in their market shares to capture variations in the extent of competition in the supply chain. More precisely, we study the case of leaders splitting in half (Leaders Split), the two largest leaders merging (Leaders Merge), Exit of largest firm, Equal market shares and a limit case of Perfect Competition. We also consider comparative static results from changes in key parameters affecting the production decision of the farmers. We explore (arbitrary) changes in international prices, production costs, endowments and risk and food security aversion. We are interested in price changes of the agricultural goods produced in Zambia. We will then use these simulated price changes with our household survey data to assess the welfare and poverty implications.

We investigate the three crops described above in the non-parametric regressions: maize, cotton and tobacco all of them exportable. Given the complexity of the scenarios, we simplify the analysis by working with a sequence of partial equilibrium models, dealing with each case separately. That is, when we take the case of cotton we leave the maize and tobacco markets unmodeled.

We start with maize. Maize is an exportable food crop by Zambia and therefore we model it as such. The top two suppliers have 22 and 20 % of the market, but the rest of the firms are small. The price changes from the simulation results are presented in Table 5.4. The first row shows the impacts of changes in competition. Given that suppliers are oligopsonies, an increase in competition should lead to an increase in prices, as it indeed happens. However, the increase is not large, only 2.34 % in the case of Perfect Competition, because of the large number of small firms.

We now focus on the role of household constraints, which are shown in Column 1 of Table 5.4, starting in row 2. This is the baseline model, different rows correspond to different comparative static results. The largest impact come from changes in international prices: if international prices increase 10 %, farmgate prices would increase by 11.98 %. This is consistent with findings in Porto et al. (2011), on which our model builds. In row 3 we simulate a 10 % increase of the Marginal cost of producing the cash crop. The effect is negative −0.27 %, since a less profitable cash crop production makes producers switch to food crops, increasing aggregate supply and therefore pushing prices down. A similar mechanic takes

Table 5.4 Farmgate price simulation results for maize

	Baseline	Leader split	Leaders merge	Exit of largest	Equal marquet shares	Perfect competition
Competition policy	0.00	0.17	−0.17	−0.19	1.33	2.34
Increase of 10 % in						
International price	11.98	12.23	11.70	11.69	13.31	15.50
Marginal cost of producing cash crop	−0.27	−0.09	−0.47	−0.49	1.06	2.34
Fixed cost of producing cash crop	−0.11	0.06	−0.29	−0.31	1.22	2.34
Endowment	−0.08	0.09	−0.25	−0.26	1.22	2.34
Preference parameter	1.08	1.18	0.99	0.97	2.47	2.34
Cash crop price	0.70	0.81	0.58	0.56	2.02	2.34
Marginal cost of producing food crop	0.54	0.67	0.41	0.39	1.83	2.34
Non-farmer demand	0.05	0.21	−0.12	−0.14	1.37	2.34

Source: Simulation results from the model of Chap. 2

place in row 4, decreasing prices by 0.11 % but shocking the Fixed Cost of cash crop instead. In the row below the shock of Fixed Costs, we show the effect on prices of an increase of 10 % in Household Resources, or, in terms of the model, the endowment. The effect is negative: −0.08 %, since an increase in resources makes households increase its production possibilities, supply more and therefore push prices down. As the last in this group we see the effects of an increase in the Risk and Food Security parameter in the 6th row. With a higher risk aversion, households produce more for autoconsumption than for the market, withdrawing supply and increasing prices 1.08 %.

In the third group of results we study the interactions between different crop markets. We start by shocking the price of the cash crop. Since the cash crop is a rival in production to maize, we would expect households to withdraw supply for maize and increase supply of cash crop, leading to higher prices for maize. This is indeed what we show in row 7: an increase of 0.70 %. An increase in marginal cost of food crop has the opposite effect of an increase of the marginal cost of cash crop mentioned above; households switch from maize to the cash crop as the former becomes unprofitable and prices increase. Finally, in the last row we show the effect of an increase of food demand, which unsurprisingly increases the price of food crop.

We move on to study the complementarities and substitutabilities. An example of the former could be the combination of Perfect competition with an increase of 10 % in International prices; the sum equals 14.32 % but the joint effect is 15.50 %. The most striking example of substitutabilities remain the combination of internal shocks with perfect competition, since perfect competition makes the model converge to a canonical model of international trade, taking away all effects of changes in internal supply and demand on prices.

We turn now to the case of Cotton and Tobacco. We report its results on Table 5.5 for cotton and Table 5.6 for tobacco. Unlike maize, these are export cash crops. The market structure of the suppliers of these crops is, like in Malawi, much more concentrated than that of maize, and even more than the market structure of the cotton suppliers in Malawi. The two largest cotton suppliers have 47 and 20 % market share, whereas for tobacco the market shares are 30 and 20 %. We should then expect larger price variations, as it is the case.

We summarize the results. In the first row we see that if the Leader would split, farmgate prices would rise 3.79 % and 2.02 % for cotton and tobacco respectively. Consistently, if Leaders Merge cotton and tobacco farmgate prices decrease 2.44 % and 1.50 % respectively. Farmgate prices for the two crops decrease again in the case of Exit of largest firm, 4.49 % for cotton and 2.82 % for tobacco. Farmgate prices increase 19.10 % for cotton and 4.63 % for tobacco if the two largest firms were to have Equal Market shares. The high increase experience by cotton prices is due to the high asymmetry in firm size: the biggest firm with 47 % market share and the second less than half of that: 20 %. Finally, were the market structure to converge to Perfect competition, farmgate prices would rise 31.52 % for cotton and 13.30 % for tobacco. In conclusion, cotton's concentration and asymmetry lead to larger results.

In Row 2 we see that for cotton, a 10 % increase in international prices leads to a 27.53 % increase in farmgate prices, whereas for Tobacco the increase in farmgate prices is 8.65 %.

Table 5.5 Farmgate price simulation results for cotton

	Baseline	Leader split	Leaders merge	Exit of largest	Equal marquet shares	Perfect competition
Competition policy	0.00	3.79	−2.44	−4.49	19.10	31.52
Increase of 10 % in						
International price	27.53	31.56	24.46	22.32	47.61	62.77
Marginal cost of producing cash crop	0.17	4.12	−2.25	−4.18	19.23	31.52
Fixed cost of producing cash crop	0.39	4.76	−1.73	−3.63	19.62	31.52
Endowment	−1.58	2.04	−4.05	−6.14	17.25	31.52
Preference parameter	2.51	8.32	0.55	−1.49	21.98	31.52
Food crop price	0.77	4.92	−1.35	−3.16	19.53	31.52
Marginal cost of producing food crop	−0.30	3.49	−2.73	−4.85	18.97	31.52
Non-farmer demand	0.00	3.79	−2.44	−4.49	19.10	31.52

Source: Simulation results from the model of Chap. 2

Table 5.6 Farmgate price simulation results for tobacco

	Baseline	Leader split	Leaders merge	Exit of largest	Equal marquet shares	Perfect competition
Competition policy	0.00	2.02	−1.50	−2.82	4.63	13.30
Increase of 10 % in						
International price	8.65	11.38	6.17	4.67	13.91	24.66
Marginal cost of producing cash crop	2.39	4.10	1.52	0.24	6.69	13.30
Fixed cost of producing cash crop	0.40	2.31	−1.11	−2.40	4.98	13.30
Endowment	−1.00	1.07	−2.57	−3.83	3.47	13.30
Preference parameter	1.79	3.71	0.55	−0.88	6.58	13.30
Food crop price	2.19	3.76	1.26	0.18	6.22	13.30
Marginal cost of producing food crop	−0.33	1.84	−2.05	−3.43	4.51	13.30
Non-farmer demand	0.00	2.02	−1.50	−2.82	4.63	13.30

Source: Simulation results from the model of Chap. 2

In Rows 3–6 we present the effects of internal shocks. An increase in the marginal cost of the cash crop leads to increase in the farmgate price of cotton and tobacco of 0.17 and 2.39 %. An increase in Fixed costs of 10 % also increases the price of the crops almost identical for both crops 0.39 % for cotton and 0.40 %

for tobacco (row 4). We see in row 5 that an increase in household resources decreases farmgate prices 1.58 and 1 % for cotton and tobacco respectively. The last of our internal shocks is an increase in risk aversion in row 6 which corresponds to price increases of 2.51 % for cotton and 1.79 for tobacco.

Finally we focus on the shocks to the alternative food crop, in rows 7–9. A rise of the food crop makes the farmgate price increase: 0.70 % in the case of cotton and 0.77 % in the case of tobacco as row 7 shows. Row 8 shows if the marginal costs of producing the food crop increase 10 %, there are decreases of 0.30 and 0.33 % for cotton and tobacco respectively. Finally, like with Malawi, an increase in non-farmer demand has no effect on the price of the cash crop.

5.4 Welfare Simulations

We complete our analysis as we did with precedent countries by discussing the welfare and poverty impacts of the comparative static results presented above, matching them with the income/expenditure shares from household surveys.

The welfare impacts of price changes are shown in Tables 5.7, 5.8 and 5.9 for maize, cotton and tobacco. We report average as well as disaggregated results for specific groups: poor/non-poor and producers.

Table 5.7 Maize price changes and household welfare

	Baseline	Leader split	Leaders merge	Exit of largest	Equal market shares	Perfect competition
Total						
Competition policy	0.00	−0.01	0.01	0.01	−0.09	−0.16
International price	−0.80	−0.82	−0.78	−0.78	−0.89	−1.03
Poor						
Competition policy	0.00	−0.01	0.01	0.01	−0.08	−0.14
International price	−0.69	−0.71	−0.68	−0.68	−0.77	−0.90
Non poor						
Competition policy	0.00	−0.01	0.01	0.01	−0.10	−0.17
International price	−0.87	−0.88	−0.85	−0.85	−0.96	−1.12
Producers						
Competition policy	0.00	0.00	0.00	0.00	0.00	−0.01
International price	−0.04	−0.04	−0.04	−0.04	−0.05	−0.05

Note: First order impact on household welfare

Table 5.8 Cotton price changes and household welfare

	Baseline	Leader split	Leaders merge	Exit of largest	Equal market shares	Perfect competition
Total						
Competition policy	0.00	0.04	−0.02	−0.04	0.18	0.29
International price	0.26	0.29	0.23	0.21	0.44	0.58
Poor						
Competition policy	0.00	0.04	−0.03	−0.05	0.20	0.32
International price	0.28	0.32	0.25	0.23	0.49	0.65
Non poor						
Competition policy	0.00	0.03	−0.02	−0.04	0.17	0.27
International price	0.24	0.27	0.21	0.19	0.41	0.54
Producers						
Competition policy	0.00	0.53	−0.34	−0.63	2.66	4.38
International price	3.83	4.39	3.40	3.10	6.62	8.73

Note: First order impact on household welfare

Row 1 shows the overall effect of competition policy. Unlike with Malawi, there are crops for which the increase of competition does not increase welfare. Namely, maize. This is a contradiction that is possibly caused by the irregular trade-patter of the crop. We have modeled maize as an export crop, as it is usually the case. But for some years, possibly due to poor harvests, imports soar and Zambia becomes a net importer. This appears be the case for the year for which we have the household data for.

In row 2 we analyze the overall effect of price increases. Again maize shows a negative sign, confirming households are, on average, net consumers of maize. This is not the case for cotton nor tobacco. Like in Malawi, maize dominates the overall welfare impact not because of its sharp price variation, but because of its larger budget shares.

Row 3 and 4 show the effect of competition policy and international price changes on the poor, and rows 5 and 6 for the non-poor. Again, three different possibilities arise. For maize, competition policy and the rise in international prices hurts both the poor and the non-poor. For cotton, the effect of competition policies and price increases welfare for both groups. And for tobacco, the poor are slightly hurt and the non-poor benefit slightly. This stresses the important of disaggregating data and studying each reform separately.

Finally, we show in rows 7 and 8 show the effects of competition policy and price increases on the welfare of producers. We see that maize producers still see a

Table 5.9 Tobacco price changes and household welfare

	Baseline	Leader split	Leaders merge	Exit of largest	Equal market shares	Perfect competition
Total						
Competition policy	0.00	0.00	0.00	0.00	0.00	0.00
International price	0.00	0.00	0.00	0.00	0.00	0.00
Poor						
Competition policy	0.00	0.00	0.00	0.01	−0.01	−0.02
International price	−0.02	−0.02	−0.01	−0.01	−0.03	−0.05
Non poor						
Competition policy	0.00	0.00	0.00	0.00	0.00	0.01
International price	0.01	0.01	0.01	0.00	0.01	0.03
Producers						
Competition policy	0.00	0.45	−0.33	−0.63	1.03	2.95
International price	1.92	2.52	1.37	1.04	3.08	5.47

Note: First order impact on household welfare

decrease in welfare when the price of the crop fall, meaning that they are still, on average, net consumers, albeit less so than overall households. Cotton producers see significant variations in welfare (4.38 % with Perfect Competition) and for tobacco, less so (2.95 %). This was to be expected given the increase in prices predicted in the section above.

References

KPMG (2015) Zambia's budget highlight 2015

Kuteya A (2012) Agricultural policies and food security challenges in Zambia. Indaba Agricultural Policy Research Institute

Kuteya A, Sitko N (2015) Creating scarcity from abundance: bumper harvests, high prices, and the role of state interventions in Zambian maize markets. Afr J Food Agric Nutr Dev 15 (4):10272–10289

Mason N, Tembo S (2015) Do input subsidy programs raise incomes and reduce poverty among smallholder farm households? Evidence from Zambia. Michigan State University, Department of Agricultural, Food, and Resource Economics

Porto G, Depetris Chauvin N, Olarreaga M (2011) Supply chains in export agriculture, competition, and poverty in Sub-Saharan Africa. CEPR Press, London

World Development Indicators (2016) The World Bank

Chapter 6
The Case of Ethiopia

This chapter studies how farm constraints and market structure affects households' welfare in Ethiopia. We describe the household survey data (Sect. 6.1), the food and cash crops institutional arrangements (Sect. 6.2) and we present the farmgate price simulation results from the model (Sect. 6.3) and the household welfare simulations (Sect. 6.4).

6.1 The Household Survey Data

We use household data from the 2000 Integrated Household Survey of Ethiopia. It is one of the largest datasets we have, with over 123,000 observations corresponding to more than 25,000 households. 68.08 % of the observations live in rural areas, whereas 86.35 % of the total population is rural. As usual in African countries, the population is young, 47.1 % being 14 years old or less, and only 3.2 % being older than 65. 50.8 % of the population is female, but only 25.7 % of households are headed by females. Household sizes are big: 5.88 people per household on average, urban households being slightly larger than rural ones (5.94 versus 5.87),

Figure 6.1 shows the distribution of income, measuring income as the (log) of per capita household expenditure (log pce). Rural and Urban distributions seem to be very similar. Given that majority of households are rural, the national distribution resembles the rural distribution more closely than the urban one.

We now turn to a description of sources of income and patterns of consumption across households. Starting with Table 6.1, we report consumption patterns for urban and rural regions. As expected, the share of auto-consumption is larger in rural households than in urban ones. The same is true for food consumption. The crops most relevant in Ethiopian budgets are Maize (11.8 % of total consumption

N. Depetris Chauvin et al., *Agricultural Supply Chains, Growth and Poverty in Sub-Saharan Africa*, Advances in African Economic, Social and Political Development, DOI 10.1007/978-3-662-53858-6_6

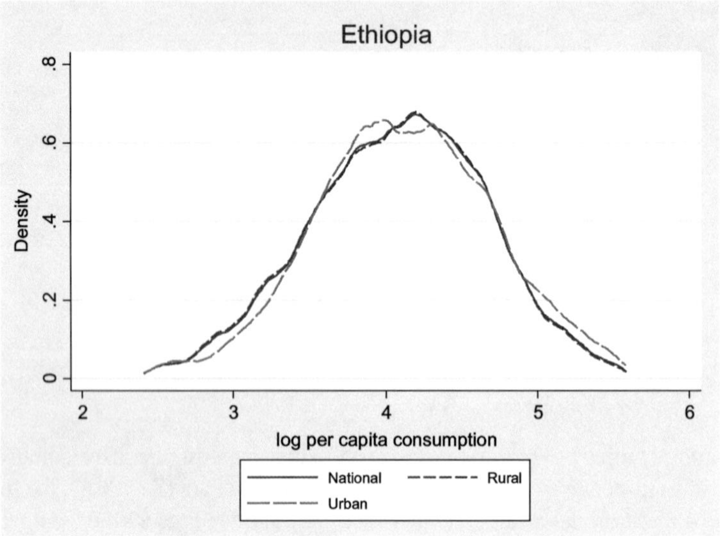

Fig. 6.1 The distribution of income density of (log) per capita household expenditure. Source: Ethiopia Integrated Household Survey (2000)

Table 6.1 Budget shares

Ethiopia	Total	Rural	Urban
Total consumption per capita	100.0	100.0	100.0
Expenditures	81.3	79.5	92.9
Food	64.1	63.7	66.8
Manufactures	8.5	8.2	10.5
Services	2.6	1.6	3.9
Others	6.0	5.9	6.7
Auto-consumption	18.7	20.5	7.1
Auto-consumption food	18.7	20.5	71
Auto-consumption others	0.0	0.0	0.0
Total food consumption	82.8	84.2	74.0
Total crops	29.9	31.8	17.5
Maize	11.8	13.0	4.4
Wheat	6.8	6.9	5.9
Sorghum	6.0	6.6	2.4
Coffee	5.2	5.3	4.7

Source: Ethiopia Integrated Household Survey (2000)

for all households, in average), Wheat (6.8 %), Sorghum (6.0 %), and Coffee (5.2 %).

In Table 6.2, we show different sources of income. An analogous phenomenon appears, rural households have lower shares of cash income than urban households (67.6 versus 82.4 %). However, the difference is much larger when it comes to food income in general (89.7 versus 24.2 %). The relative importance of crops in income

Table 6.2 Income shares

Ethiopia	Total	Rural	Urban
Total income per capita	100.0	100.0	100.0
Incomes	69.5	67.6	82.4
Food (agriculture)	50.8	57.3	6.6
Wage	7.1	2.7	36.8
Enterprises	9.6	6.0	34.2
Transfers	2.0	1.6	4.7
Auto-consumption	30.5	32.4	17.6
Auto-consumption food	30.5	32.4	17.6
Auto-consumption others	0.0	0.0	0.0
Total food income and AC	81.3	89.7	24.2
Total crops	22.5	24.7	7.0
Maize	8.5	9.5	2.0
Wheat	5.1	6.4	3.6
Sorghum	2.0	2.2	0.4
Coffee	5.9	6.6	0.9

Source: Ethiopia Integrated Household Survey (2000)

is different than in consumption. Coffee becomes significantly more relevant than sorghum, which is expected, because Coffee is a cash crop. The ordering for the rest of the crops remains unchanged: Maize (8.5 % of income on average for all households), Wheat (6.1 %), Coffee (5.9 %) and Sorghum (2.0 %).

We now focus on our graphical analysis. We plot estimates of non-parametric regressions of income and budget shares on log pce. We begin in Figs. 6.2 and 6.3 with average share of food expenditure (cash expenditure plus auto-consumption) and share of food income (cash agricultural plus auto-consumption). For rural households, the share of food consumption is generally high but drops suddenly for the wealthiest households. For urban households, the fall is more monotonic, starting from a lower level of income. It must be noted that food consumption remains very high for all households, even for the wealthiest (urban) households. We now look at the other opposite side: Figure 6.3 shows shares of food production on the logarithm of household per capita expenditure. The graph resembles the previous one only in that urban households have lower shares than rural ones. However, for each group, the share of income derived from food income remains unchanged as income increases. Apparently, wealthier households do not shift to wage income or cash-crop income (at least not to a significant degree in the aggregate).

We now disaggregate the data and look at each crop in particular (Figs. 6.4, 6.5, 6.6, 6.7, 6.8, 6.9, 6.10, 6.11 and 6.12). Starting with maize, both for expenditures and incomes, urban households are below rural ones. However, maize budget shares (Fig. 6.4) are decreasing in income for both type of households, whereas income shares (Fig. 6.5) have an inverted u-shape. Sorghum's budget shares (Fig. 6.6) and income shares (Fig. 6.7) follow the exact same patterns. Teff is completely different. Firstly, it is teff's budget share (Fig. 6.8) which follows an

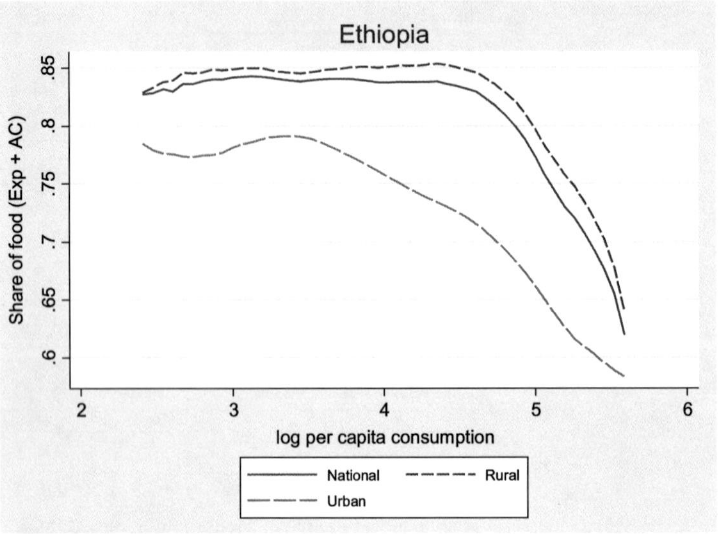

Fig. 6.2 Total food budget share across the income distribution. Source: Ethiopia Integrated Household Survey (2000)

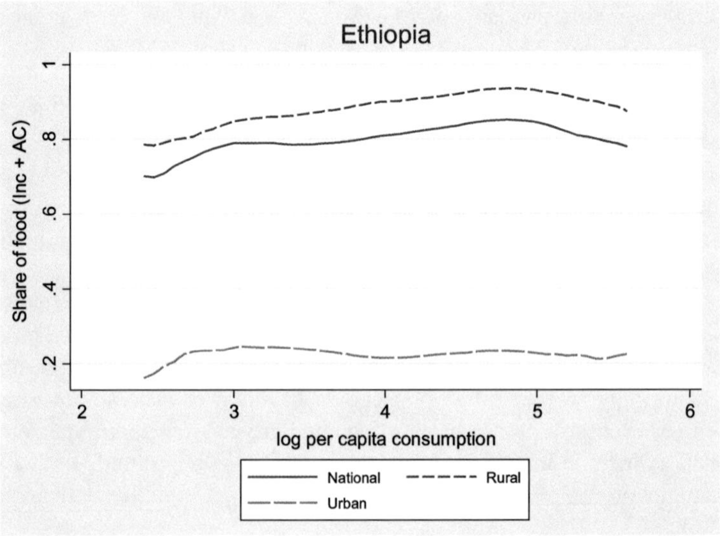

Fig. 6.3 Total food income share across the income distribution. Source: Ethiopia Integrated Household Survey (2000)

inverted u-shape. Furthermore, urban households have a larger budget share for teff than rural ones. When it comes to income shares (Fig. 6.9), teff is increasing in income, resembling some cash crops, although it is a food crop. In this case, urban

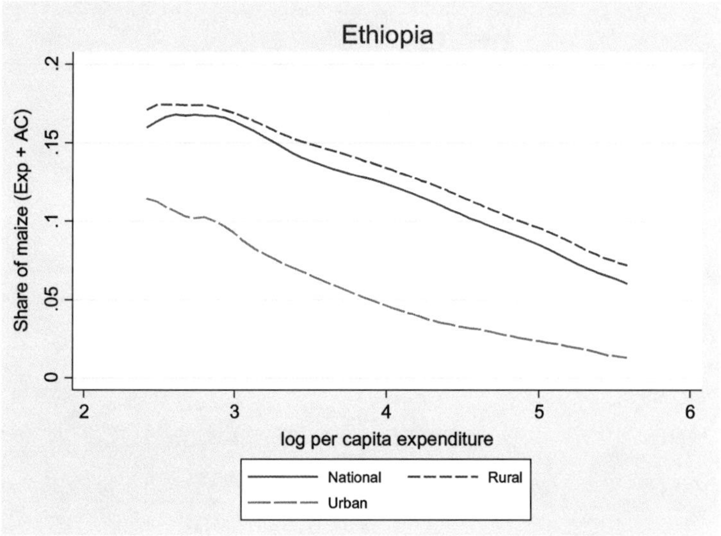

Fig. 6.4 Maize budget share across the income distribution. Source: Ethiopia Integrated Household Survey (2000)

shares have lower shares than rural shares, as usual. In Figs. 6.10 and 6.11 we show the budget and income shares for wheat, respectively. Wheat's budget shares follow a usual pattern. For rural Households, they are increasing in income, to then fall slightly for wealthier households. For urban household the share has an inverted u shape. Poorer urban households spend more in wheat than rural households with the same level of income on average, but the opposite is true for wealthier urban and rural households. Wheat's income shares follow a more common pattern: for rural households they increase monotonically on income to fall slightly for wealthier households, whereas for urban ones they increase only initially, to stagnate later. Finally, in Fig. 6.12 we show the income shares for coffee, and not the budget shares because it is a cash crop. Shares are almost zero for urban households, and follow a U shaped patter for rural ones.

6.2 Food and Cash Crops in Ethiopia

Ethiopia has great potential for agricultural development. The agricultural sector employs about 73 % of its population. In 2015, the sector recorded 41 % of GDP and an annual growth rate of 6.4 % (World Development Indicators 2016). It has an agricultural land area of 362,590 km^2, equivalent to 36.3 % of the total land area. Food exports and imports (as a percentage of merchandize exports and imports) reached 71.7 and 10.6 respectively.

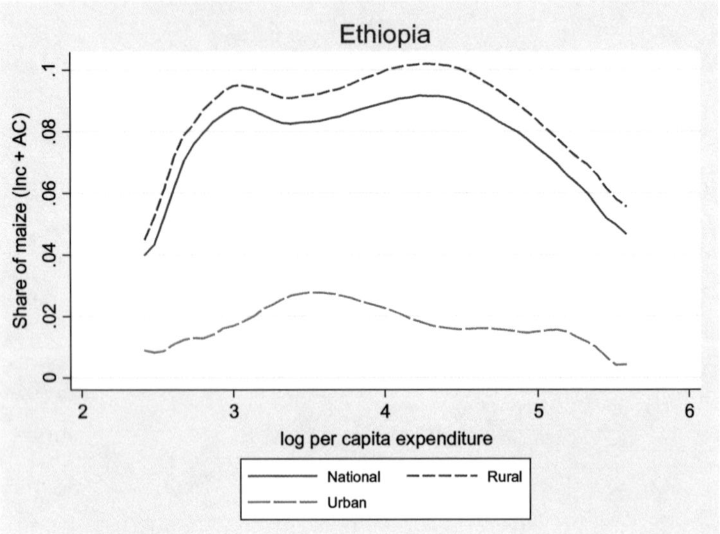

Fig. 6.5 Maize income share across the income distribution. Source: Ethiopia Integrated Household Survey (2000)

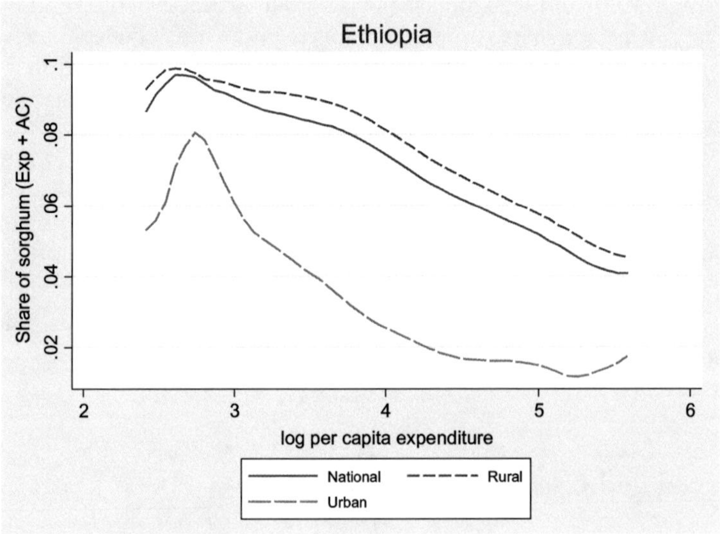

Fig. 6.6 Sorghum budget share across the income distribution. Source: Ethiopia Integrated Household Survey (2000)

The agricultural sector has low levels of technology development: fertilizer use and irrigation are minimal and there is heavy reliance on smallholder production. Most agricultural production takes place in the highlands of Ethiopia, where five

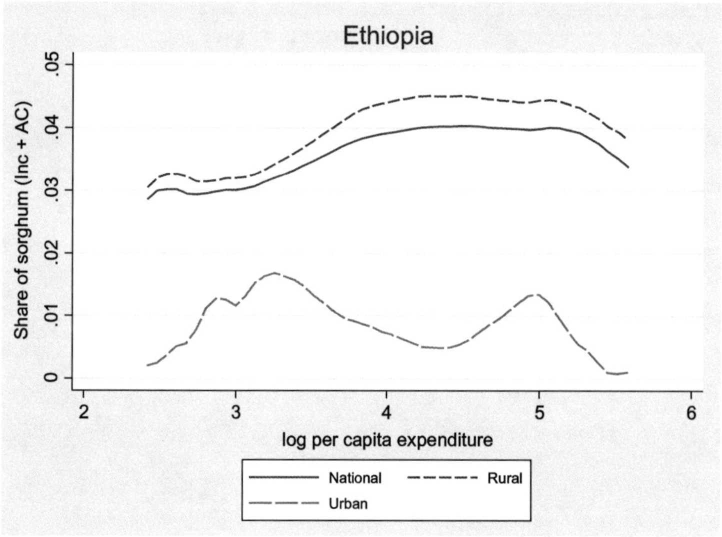

Fig. 6.7 Sorghum income share across the income distribution. Source: Ethiopia Integrated Household Survey (2000)

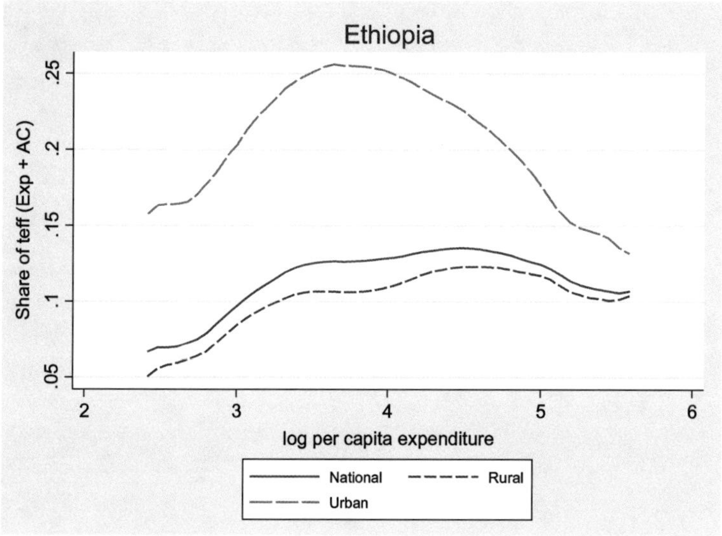

Fig. 6.8 Teff budget share across the income distribution. Source: Ethiopia Integrated Household Survey (2000)

major cereals, teff, wheat, maize, sorghum and barley are major staples and play dominant roles in cropping systems depending on elevation, rainfall, and market access. Total cereal production reached an estimated 23,607,662 mt, with yields at

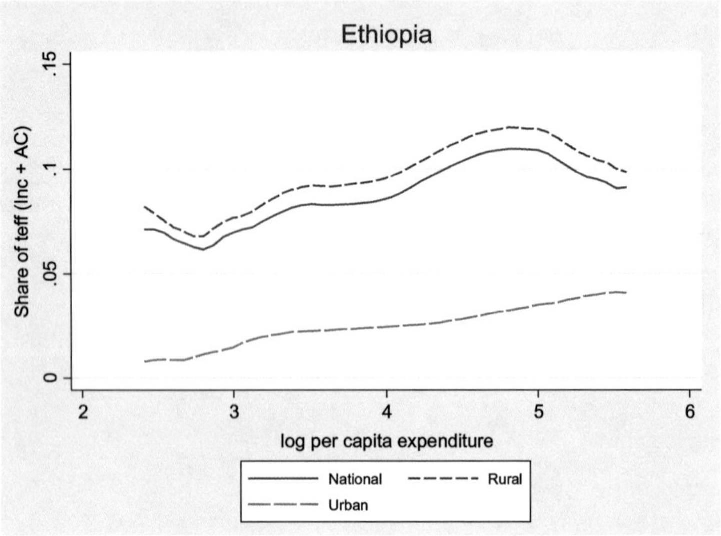

Fig. 6.9 Teff income share across the income distribution. Source: Ethiopia Integrated Household Survey (2000)

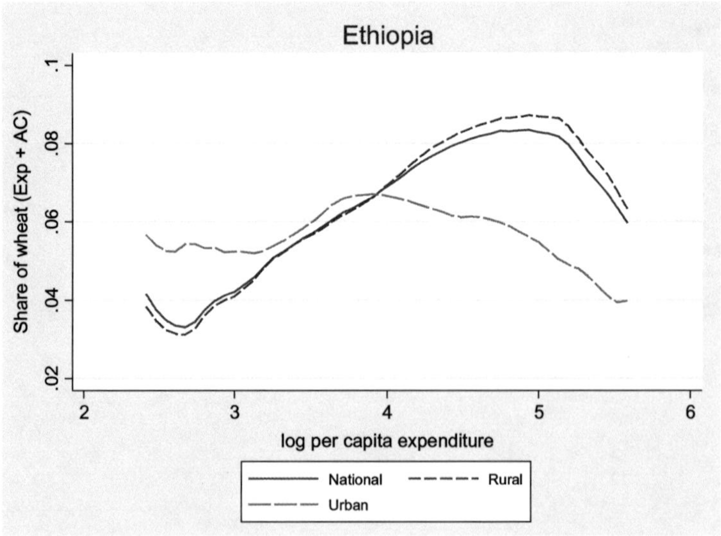

Fig. 6.10 Wheat budget share across the income distribution. Source: Ethiopia Integrated Household Survey (2000)

2325 kg/ha in 2014. For rural areas as a whole, wheat, maize, and sorghum each account for about 10 % of total food expenditures for both the poor and non-poor, with substantial regional variation. In contrast, teff accounts for about 15 % of total

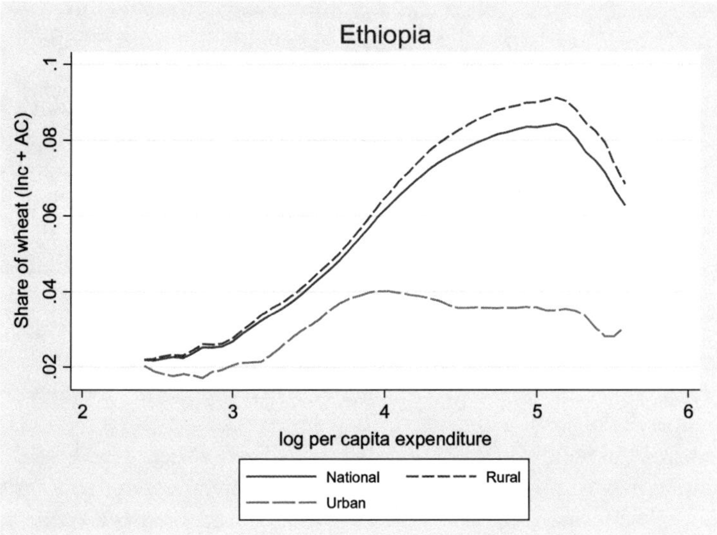

Fig. 6.11 Wheat budget share across the income distribution. Source: Ethiopia Integrated Household Survey (2000)

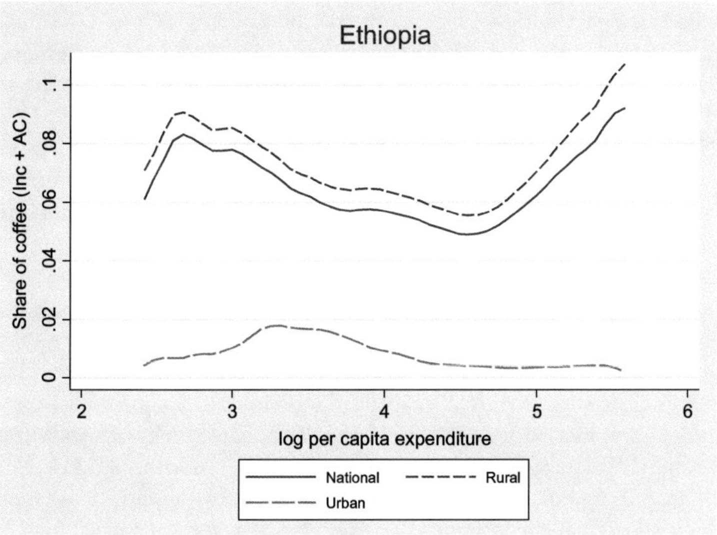

Fig. 6.12 Coffee income share across the income distribution. Source: Ethiopia Integrated Household Survey (2000)

food expenditures in urban areas, with more than 10 % of cereal expenditures on processed cereals mainly flour. In total, the five staple cereals account for two-thirds of total cultivated area and over 30 % of agricultural GDP. Pulse and oilseed follows the cereals closely as second and third most cultivated crop respectively. Coffee is the major cash crop produced, accounting for about 4 % of agricultural GDP and 3 % of total area cultivated. The government of Ethiopia places high priority on agricultural development. It has been reported that Ethiopia has the highest agricultural budget expenditure share in Africa (Headey et al. 2014). The Government strategy towards agriculture is geared towards improving productivity of smallholder farmers. In 2008, smallholder farmers cultivated 12 million hectares of land, which represent 96.3 % of the total cultivated area (Taffesse et al. 2012). To increase smallholder productivity, the government has enacted ambitious plans to develop and extend new seeds, chemical fertilizers, new crops, and new natural resource management practices like irrigation. But the government still formally owns and regulates the distribution and leasing of land, and has promoted controversial smallholder resettlement schemes and investments in large commercial farms (Headey et al. 2014).

In terms of the policy environment, there has been a focus on liberalization of agricultural markets, investment in transportation infrastructure, and agricultural research and extension (Aguilar et al. 2015). However, the market liberalization efforts by the Government has not completely left the agricultural sector to market forces. The seed and fertilizer markets are still subject to government intervention (Dorosh and Radish 2012). Central to the agricultural policy of the government is the Agricultural Growth Program, which is the agricultural-sector program for the Growth and Transformation Plan (GTP) of the country. It seeks to sustainably improve crop production and productivity through increased use by farmers of best practices and increasing the availability and adoption of improved inputs; seed and fertilizer. Cereals are at the center of most of these policy efforts. The Productive Safety Net Program (PSNP), the Land Certification Program, and the Participatory Demonstration and Training Extension System (PADETES) are prominent interventions from the Government. The PSNP implemented in 2005, aims to assure food consumption and prevent asset depletion for rural food-insecure households in a way that stimulates markets, improves access to services and natural resources, and rehabilitates and enhances the natural environment (FSCD 2007). Essentially, the program provides food and cash payments to households suffering from food insecurity for a 6 months period, and in return these households volunteer labor that builds public infrastructure. In 2005, the PSNP provided support to five million people, this has increased to eight million in 2015. The annual expenditures of the PSNP is approximately US$750 million, and is the second largest safety net program in Sub-Saharan Africa. Clarke et al. (2016) report that independent evaluations show that the program has significantly improved the food security status of beneficiaries in Ethiopia. The land certification program was fully implemented in all the regions of Ethiopia by 2007. By 2007, it was estimated that over five million households have certified their land holdings in all four regions of Ethiopia (Deininger et al. 2011). The program provides households with a document which typically includes the names and photographs of the household head and spouse, the size and location of the land holding, as well as the neighbors of the demarcated land of the households. Impact evaluation of the program indicate that it has boosted farmers' perception of tenure security, increased agricultural productivity and land related investments, as well as improved land rental market participation of female headed farm households (Deininger et al. 2011; Bezabih et al. 2011).

We have selected to analyze five crops for the case of Ethiopia. These include teff, wheat, coffee, maize, and sorghum (Figs. 6.13 and 6.14). We begin first with teff. Teff is an indigenous crop widely grown only in Ethiopia and Eritrea. Teff is used to make 'injera' (the sourdough flatbread). It is a small grain with health benefits including being gluten free. It is usually grown in the north central region of Ethiopia. Teff is the most important crop in Ethiopia, in 2011/2012, it was

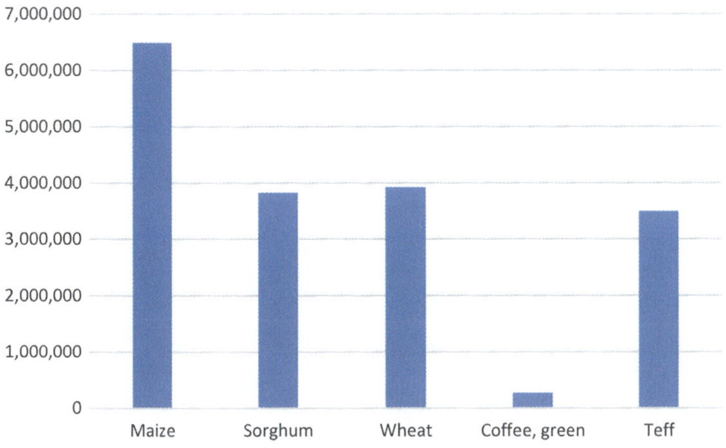

Fig. 6.13 Crops production in Ethiopia (in tons, 2013). Source: FAOSTAT Online

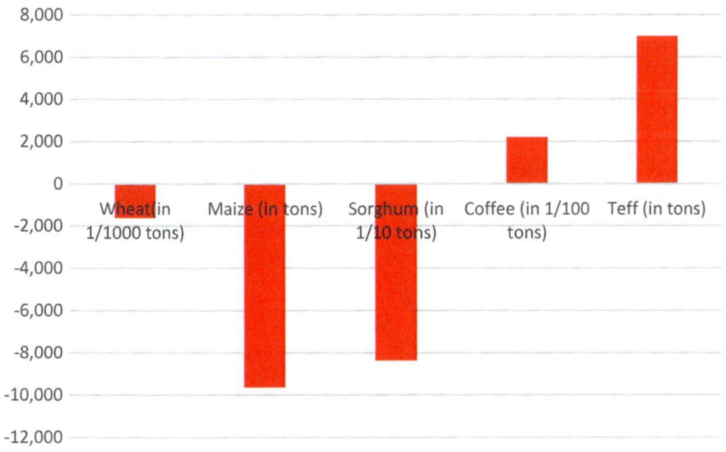

Fig. 6.14 Crops production in Ethiopia (in 2013). Source: FAOSTAT Online

estimated that teff made up 20 % of all the cultivated area in Ethiopia, covering about 2.7 million hectares and grown by 6.3 million households (CSA 2012). It accounts for between 11 and 15 % of all calories consumed in Ethiopia, and provides about 66 % of the daily protein intake. In urban areas, the share of per capita teff consumption in total food expenditure is 23 %, while in rural area this is only 6 % (Vandercasteelen et al. 2013). Urbanization has been touted as a driver of this dynamics. In 2013, the average teff yield reached 1.4 t/ha, an increase of 8 % from 2012 (CSA 2013). The increase was mainly due to an increase in production area. The value of its commercial surplus is second only to coffee. In addition, teff straw is important for fodder and used in construction. Ethiopia is a net exporter of teff. Aside from it being a low risk crop, teff production faces constraints. Yields are unusually low. Teff productivity reached 1.4 t/ha, compared to other cereals such as maize (3.1 t/ha), rice (2.8 t/ha) and wheat (2.1 t/ha) (CSA 2013). Low use of fertilizer and improved seeds account for this situation. National estimates show that only 2 % of teff farmers used improved seeds, although more than one third applied fertilizer for teff production (CSA 2012). There are also issues of erosion, low soil fertility, and significant post-harvest and processing losses.

Wheat is one of the major crops in Ethiopia in terms of cultivation and consumption. It has a volume of production at 4,231,589 mt, accounting for 18 % of total cereal production in 2014. In 2014, harvested area of wheat reached 1,663,845 ha, with yields of 2.54 t/ha. While the country is considered as the largest wheat producer in Sub-Saharan Africa, it is also a net importer of wheat, and imports occur especially during drought years. Smallholder wheat production is prominent, allowing it to meet more than 70 % of the demand from domestic production (Shiferaw et al. 2011). Of the total wheat production area, 76 % is located in Bale, Arsi and Shewa regions. In Ethiopia, both durum wheat and bread wheat are widely cultivated. While bread wheat is a recent introduction to Ethiopia, durum wheat is indigenous and mainly grown in the Central and Northern highlands. There are about 4.6 million farm households, which is about 36 % of cereal farm households who are directly dependent on wheat farming in Ethiopia. Despite the low yields, demand for wheat has been growing fast in both rural and urban areas in the country. This has been attributed to increasing incomes and urbanization. Wheat production faces many constraints. Principal among them is low yields. The national average productivity of wheat is well below the experimental yield of above 5 t/ha. In 2012, Ethiopia's wheat yield was 29 % below the Kenyan average, 13 % below the African average, and 32 % below the world average (FAOSTAT Online). There are also issues of low seed varieties and low fertilizer application. Only about 1 % of the wheat area was cultivated using improved seed—fertilizer package (CSA 2012).

Coffee is the major export crop in Ethiopia, accounting for 22 % of Ethiopia's commodity exports in 2013/2014 (NBE 2014). Ethiopia is a net exporter of coffee. Also, Ethiopia is the biggest coffee exporter in Africa, accounting for 3 % of the global coffee trade (ICO 2014). It is estimated that coffee is cultivated by over four million smallholder farming households in Ethiopia (CSA 2013). It had a volume of production at 270,000 mt in 2013. In 2013, harvested area of coffee reached

520,000 ha, with yields of 519.2 kg/ha. In global markets, Ethiopian coffee is valuable because it is of the Arabica type, and because of its unique taste. There is high level of consumption of coffee within Ethiopia, usually about half of the production. Trading of coffee by private traders and rural processor is done through the Ethiopian Commodity Exchange (ECX), after the government enacted a law for more transparent trading. The cash crop is not without constraints. Coffee yields are relatively low. Compared to other coffee producing African countries, Ethiopian yields, at 450 kg/ha are slightly higher than in Kenya (280 kg/ha), and Rwanda (410 kg/ha), but lower than in Uganda (600 kg/ha). Compared with the major Latin American producers Brazil, Guatemala and Costa Rica, Ethiopian yields are only one-half to one-third of the level achieved in these countries. Also, Ethiopia's farmers obtain a smaller share of export prices compared to most other countries. In comparison to four other Arabica—producing countries, Ethiopian farmers earn the lowest share of the export price, at 60 % of the export value, with shares in other countries ranging from 70 % in Kenya to 90 % in Brazil (Minten et al. 2015).

Maize continues to be a significant contributor to the economic and social development of Ethiopia. With a total production of 6.5 million tons in 2013, it is the most produced crop in Ethiopia in terms of quantity. In addition to the highest total production per annum and the highest per-hectare yield, maize is also the single most important crop in terms of number of farmers engaged in its cultivation. Maize is also one of the most consumed crop in Ethiopia. A study conducted by Rashid et al. (2010) stated that the average Ethiopian consumes a total of 1858 kcal daily of which four major cereals (maize, teff, wheat, and sorghum) account for more than 60 %, with maize and wheat representing 20 % each. Maize is grown primarily in the Amhara, Oromia and SNNP regions of Ethiopia. Despite its prominence, it is important to mention that maize farmers in Ethiopia face a series of challenges that limit their overall production and income. The key challenges can be broadly categorized into three groups: (i) lower yields due to limited use of modern inputs; (ii) majority of sales immediately after harvest; and (iii) high post-harvest losses (both on- and off-farm).

The drought tolerance and adaptation attributes of sorghum has made it the favorite crop in drier and marginal areas of Ethiopia. That is why sorghum is cultivated by nearly 4.5 million smallholders located in the eastern and northwest parts of the country, where the weather is dry and soil fertility is poor. Specifically, the main sorghum producing regions are Oromia and Amhara, accounting for nearly 80 % of the total production. The leading sorghum producing zones are East and West Hararge in Oromiya and North Gondar and North Shoa in Amhara. Two regions, SNNPR and Tigray, are relatively less important, contributing 11 and 4 % of the national production, respectively. Sorghum production was nearly 3.8 million tons in 2013, making Ethiopia Africa's second largest sorghum producer. However, the crop faces a number of challenges. In the forefront we have the poor marketing system. Demeke and Marcantonio (2013) argue that only 11.5 % of the crop is sold with 74.0 % being consumed at the local level. This is why Ethiopia is a net sorghum importer. While the introduction of the Ethiopia Commodity Exchange in 2008

Table 6.3 Market shares in Ethiopia

Coffee		Wheat	
Company	Shares (%)	Company	Shares (%)
Ethiopian Grain Trade Enterprise	5.50	Hafiya Flour Factory	6.43
Muluneh Kaka Dumeso	5.60	DireDawa Food Complex	3.11
Oromia Coffee Farmers Coop P. LC	3.80	Universal Food Complex Flour Factory	2.33
Tracon Trading PLC	4.10	Juked Flour Factory	2.33
Cabey Private Limited Company	4.10	Ayan Fuad Flour Factory	1.98
Alfoz PLC	4.20	Africa Flour Factory	1.81
A.T.L Trading PLC	4.20	Haji Workicho	1.59
S.A. Bagersh PLC	3.50	D.H Geda Flour Factory	1.56
GMT Industrial Private Limited Comp	3.20	Kebron Flour Factory	1.56
Adem Kedir Hajihassen	3.00	Jemaneh Flour Factory	1.56
Mullege Private Limited Company	2.10	AdmasTesfa	1.52
Sidama Coffee Farmers Coopera Union	2.10	Muna	1.49
Others	54.60	Others	72.73

Source: Stakeholders interviews

sought to provide the necessary infrastructure to help improve sorghum and other crops marketing, the poor yields at 2.1 t/ha, remains one of the main constraints.

Table 6.3 below displays the market shares for processors of wheat and coffee in Ethiopia. Coffee and wheat are crops traded on the Ethiopian Commodity Exchange (ECX), and their price is determined depending on market forces influenced by international market price. The influence of ECX on wheat is rather minimal. For formal milling of wheat, Hafiya Flour factory processes about 10,746 t/month and has a market share of 6.43 %, while DireDawa factory, Universal factory, Juked flour factory, and Ayan factory processes 5200, 3900, 3900, and 3302 t/month respectively, with market shares of 3.11, 2.33, 2.33, and 1.98 %. The rest (top 32 out of 146) have control a market share of 34.41 %. Marketing of coffee is done by various players. Muluneh Dumeso and the Ethiopian Grain trading enterprise control a market share of 5.6 and 5.5 % respectively. The rest of the market is controlled by various players. For the market shares of teff, we do not have much detail as there are many wholesalers. Teff has relatively larger share of marketable surplus compared to wheat and coffee but thousands of small traders engage in its purchase and sales without any official record about their share. We will assume a limit model for perfect competition. The prices of Teff are determined by market forces without any government influence.

The World Food Program (WFP) is a major purchaser of maize for its food-aid intervention from major maize producing areas of the country though Ethiopian Grain

Trade Enterprise (EGTE) and Farmer's Cooperatives based on its new project called "Purchase for Progress". The total purchase through its agents in 2010/2011 was 19,000 t. In the same year Ethiopia's maize marketable surplus was estimated to be 568,418 t. WFP local purchase as share of surplus production is 3.34 %. Donors—are the second biggest purchasers of maize from surplus regions and distribute as food-aid in the deficit regions. There is no organized data to account their share. The Disaster Risk Management and Food Security (DRMFS) which is the government body purchases maize for food-reserve to be used when there is an emergency. Both the shares of donors and DRMFS are small. The remaining market shares is shared among a large number of small wholesalers. Last, the market share of sorghum can be considered as very competitive. There is a large number of small traders who purchase from surplus areas and supply to deficit regions. But their share is not known.

6.3 Simulation Results

In this section, we use the model again to perform various simulations. This time we investigate five case studies: coffee (exportable cash crop), teff (exportable food crop), maize, sorghum and wheat (food importables). We deal with each crop separately in a partial equilibrium model, that is, when analyzing one crop we leave the rest of the markets unmodeled. It must be noted that for all of the crops under study the supply chain is very competitive. This will inevitably lead to small price changes when we alter market structure.

We start by describing the results for coffee. As we mentioned earlier, coffee is a cash crop exported by Ethiopia and we therefore use the cash crop export model. We present the price changes from the simulation results in Table 6.4. Unsurprisingly, the extent of competition leads to small price changes in the simulations. For the changes in market structure (Row 1), a change towards Perfect Competition, for instance, achieves only a slight price increase of 0.98 %. In Row 2, however, large elasticities in arise with regards to an increase international prices: local prices increase 30.35 % when international prices increase 10 %. The rest of the internal shocks presented in Rows 3–9 have the expected sign and magnitude, none of them being very large, for the reasons we explained before. When it comes to complementarities we can note the combination of international price increase and convergence to Perfect Competition, being the sum of the effects 31.33 % and the global effect 32.33 %. An example of substitutabilities would be a convergence to Perfect Competition and an increase in the marginal cost of the cash crop, being the sum of the effects 2.34 % and the global effect 0.98 %. Given the small magnitude of the results, few other notable complementarities and substitutabilities arise.

Our next crop in the list is teff, which is an exportable food crop. Its supply chain is also very competitive. We should not expect the same results, however, because the rest of the parameters are different. Nevertheless, since market structure is the most relevant parameter in our model, results are unlikely to be too different. Indeed, in Table 6.5 the first row shows that changes in market structure bring

Table 6.4 Farmgate price simulation results for coffee

	Baseline	Leader split	Leaders merge	Exit of largest	Equal market shares	Perfect competition
Competition policy	0.00	0.05	−0.04	−0.04	0.00	0.98
Increase of 10 % in						
International price	30.35	30.44	30.25	30.25	30.35	32.23
Marginal cost of producing cash crop	1.36	1.34	1.34	1.34	1.34	0.98
Fixed cost of producing cash crop	0.07	0.11	0.02	0.02	0.07	0.98
Endowment	−0.18	−0.13	−0.23	−0.23	−0.18	0.98
Preference parameter	0.31	0.34	0.27	0.27	0.31	0.98
Food crop price	0.95	0.95	0.95	0.95	0.95	0.98
Marginal cost of producing food crop	−0.15	−0.09	−0.20	−0.20	−0.15	0.98
Non-farmer demand	0.00	0.05	−0.04	−0.04	0.00	0.98

Source: Simulation results from the model of Chap. 2

Table 6.5 Farmgate price simulation results for teff

	Baseline	Leader split	Leaders merge	Exit of largest	Equal market shares	Perfect competition
Competition policy	0.00	0.02	−0.03	−0.03	0.00	0.57
Increase of 10 % in						
International price	22.29	22.36	22.21	22.21	22.29	23.82
Marginal cost of producing cash crop	−0.41	−0.36	−0.45	−0.45	−0.41	0.57
Fixed cost of producing cash crop	−0.12	−0.09	−0.15	−0.15	−0.12	0.57
Endowment	−0.01	0.02	−0.04	−0.04	−0.01	0.57
Preference parameter	1.04	1.02	1.07	1.07	1.04	0.57
Cash crop price	0.76	0.76	0.77	0.77	0.76	0.57
Marginal cost of producing food crop	0.19	0.21	0.18	0.18	0.19	0.57
Non-farmer demand	0.03	0.05	0.00	0.00	0.03	0.57

Source: simulation results from the model of Chap. 2

small effects, even smaller than those for Coffee. If the supply chain were to become perfectly competitive, prices would increase only by 0.57 %. In Row 2 we see that the elasticity of internal prices with respect to international prices decreases, but remains high: a 10 % increase in international prices leads to a 22.29 % rise in internal prices. The rest of the results presented in Rows 3–9 are like

Table 6.6 Farmgate price simulation results for maize

	Baseline	Leader split	Leaders merge	Exit of largest	Equal marquet shares	Perfect competition
Competition policy	0.00	−0.07	0.08	0.08	0.00	−1.75
Increase of 10 % in						
International price	3.79	3.77	3.83	3.83	3.79	2.88
Marginal cost of producing cash crop	−0.63	−0.67	−0.57	−0.57	−0.63	−1.75
Fixed cost of producing cash crop	−0.76	−0.81	−0.72	−0.72	−0.76	−1.75
Endowment	0.92	0.82	1.03	1.03	0.92	−1.75
Preference parameter	1.19	1.08	1.31	1.31	1.19	−1.75
Cash crop price	2.04	1.89	2.21	2.21	2.04	−1.75
Marginal cost of producing food crop	1.00	0.89	1.11	1.11	1.00	−1.75
Non-farmer demand	0.03	−0.04	0.12	0.12	0.03	−1.75

Source: Simulation results from the model of Chap. 2

in most of our simulations, very small. Complementarities arise in the usual combinations: international prices and Perfect Competition (global effect 23.82 %, sum of effects 22.86 %), and so do substitutabilities: Perfect Competition with, for example, an increase of the Risk Parameter (global effect 0.57 %, sum of individual effects 1.32 %).

The third crop under analysis is maize, an importable food crop. Its simulation results are shown in Table 6.6. Given that maize has a market structure similar to coffee and teff but is an importable crop, we should expect similar results but with a sign reversal. This is exactly what happens: we see in Row 1 that the strongest effect is as always a change towards Perfect Competition, which amounts this time to a decrease of 1.75 % in local prices. Row 2 shows very low elasticities of internal prices with respect to international prices: an increase of 10 % in international prices leads to an increase of only 3.79 % in local prices. This is not much larger than the effect of an increase of 10 % in the Risk and Food Security Parameter in Row 6 (1.19 %) and in the price of the Cash Crop coffee (2.04 %). An example of a complementarity would be Perfect Competition paired with an increase in the price of a Cash Crop (global effect −1.75 %, sum of effects 0.28 %), substitutabilities arise in for example Perfect competition paired with an increase in the Fixed Costs (global −1.75, −2.52 %).

The fourth crop we simulate is sorghum, which we present in Table 6.7. Similar results arise. In Row 1 we see that the largest effect is caused by making the supply chain perfectly competitive, but the decrease in prices is only 0.90 %. In Row 2 we see that the elasticity of local prices with respect to international prices is larger than in maize: an increase of 10 % in international prices leads to an increase of 6.58 % in local prices, ceteris paribus. The rest of the results presented in Rows 3–9

Table 6.7 Farmgate price simulation results for sorghum

	Baseline	Leader split	Leaders merge	Exit of largest	Equal marquet shares	Perfect competition
Competition policy	0.00	−0.04	0.04	0.04	0.00	−0.90
Increase of 10 % in						
International price	6.58	6.57	6.59	6.59	6.58	6.35
Marginal cost of producing cash crop	−0.11	−0.14	−0.07	−0.07	−0.11	−0.90
Fixed cost of producing cash crop	−0.47	−0.48	−0.44	−0.44	−0.47	−0.90
Endowment	0.58	0.51	0.64	0.64	0.58	−0.90
Preference parameter	0.86	0.78	0.94	0.94	0.86	−0.90
Cash crop price	1.10	1.01	1.20	1.20	1.10	−0.90
Marginal cost of producing food crop	0.37	0.32	0.43	0.43	0.37	−0.90
Non-farmer demand	0.02	−0.02	0.06	0.06	0.02	−0.90

Source: Simulation results from the model of Chap. 2

are quite small, the largest one being an increase in the price of the rival crop, the Cash Crop in Row 7 with (1.10 %). There is a complementarity in the joint effect of Perfect Competition and an increase of the Food and Risk Parameter (joint effect is −0.90 % and the sum is 0.04 %), for example, and a substitutability in the joint effect of Perfect Competition and the Fixed Cost of producing the Cash Crop (global −0.90 % and sum −1.37 %).

The fifth and last crop we analyze is wheat, presented in Table 6.8. As expected, changes in market structure, presented in Row 1, have small effects. A change towards Perfect Competition makes prices decrease by 2.62 %. In Row 2 we see that the elasticities of internal prices with respect international price are slightly above those for Maize: an increase of international prices of 10 % leads to a rise in farmgate prices of 4.21 % ceteris paribus. From rows 3–9, the effects of an increase in the price of the Cash Crop (3.28 %) and an increase in the Marginal Cost of Producing the Crop stand out (−1.66 %). Precisely, pairing them with Perfect Competition leads to the largest complementarity in the case of an increase of the Price of the Cash Crop (global: −2.62 % and the sum 0.66 %) and the largest substitutability, with the increase of the Marginal Cost of Producing the Crop (global again −2.62 % and sum −4.28 %).

Table 6.8 Farmgate price simulation results for wheat

	Baseline	Leader split	Leaders merge	Exit of largest	Equal marquet shares	Perfect competition
Competition policy	0.00	−0.11	0.13	0.13	0.00	−2.62
Increase of 10 % in						
International price	4.21	4.11	4.31	4.31	4.21	1.95
Marginal cost of producing cash crop	−1.66	−1.71	−1.62	−1.62	−1.66	−2.62
Fixed cost of producing cash crop	−0.52	−0.61	−0.42	−0.42	−0.52	−2.62
Endowment	0.67	0.53	0.82	0.82	0.67	−2.62
Preference parameter	−0.60	−0.69	−0.49	−0.49	−0.60	−2.62
Cash crop price	3.28	3.03	3.56	3.56	3.28	−2.62
Marginal cost of producing food crop	0.14	0.02	0.28	0.28	0.14	−2.62
Non-farmer demand	0.09	−0.03	0.22	0.22	0.09	−2.62

Source: Simulation results from the model of Chap. 2

Table 6.9 Coffee price changes and household welfare

	Baseline	Leader split	Leaders merge	Exit of largest	Equal market shares	Perfect competition
Total						
Competition policy	0.00	0.00	0.00	0.00	0.00	0.02
International price	0.67	0.67	0.67	0.67	0.67	0.71
Poor						
Competition policy	0.00	0.00	0.00	0.00	0.00	0.02
International price	0.72	0.72	0.72	0.72	0.72	0.76
Non poor						
Competition policy	0.00	0.00	0.00	0.00	0.00	0.02
International price	0.63	0.64	0.63	0.63	0.63	0.67
Producers						
Competition policy	0.00	0.01	−0.01	−0.01	0.00	0.15
International price	4.70	4.72	4.69	4.69	4.70	4.99

Note: First order impact on household welfare

Table 6.10 Teff price changes and household welfare

	Baseline	Leader split	Leaders merge	Exit of largest	Equal market shares	Perfect competition
Total						
Competition policy	0.00	0.00	0.00	0.00	0.00	0.01
International price	0.51	0.52	0.51	0.51	0.51	0.55
Poor						
Competition policy	0.00	0.00	0.00	0.00	0.00	0.01
International price	0.55	0.55	0.54	0.54	0.55	0.5 S
Non poor						
Competition policy	0.00	0.00	0.00	0.00	0.00	0.01
International price	0.55	0.55	0.54	0.54	0.55	0.58
Producers						
Competition policy	0.00	0.00	0.00	0.00	0.00	0.06
International price	2.30	2.30	2.29	2.29	2.30	2.45

Note: First order impact on household welfare

Table 6.11 Maize price changes and household welfare

	Baseline	Leader Split	Leaders merge	Exit of largest	Equal market shares	Perfect Competition
Total						
Competition policy	0.00	0.01	−0.01	−0.01	0.00	0.13
International price	−0.28	−0.27	−0.28	−0.28	−0.28	−0.21
Poor						
Competition policy	0.00	0.01	−0.01	−0.01	0.00	0.17
International price	−0.33	−0.37	−0.33	−0.38	−0.33	−0.29
Nor poor						
Competition policy	0.00	0.00	0.00	0.00	0.00	0.10
International price	−0.21	−0.21	−0.21	−0.21	−0.21	−0.16
Producer						
Competition policy	0.00	0.00	−0.01	−0.01	0.00	0.12
International price	−0.27	−0.27	−0.27	−0.27	−0.27	−0.20

Note: First order impact on household welfare

6.4 Welfare Simulations

We complete our analysis of the crops in Ethiopia by discussing the welfare and poverty impacts of the results presented in the previous section. The welfare impacts of the price changes are reported in Tables 6.9, 6.10, 6.11, 6.12 and 6.13 for coffee, teff, maize, sorghum and wheat.

Some regularities are worth mentioning. As always, every increase in competition leads to increase in welfare, and a rise in international prices leads to an increase in welfare for exportable crops and a reduction for importable crops. The small price variations from the previous section lead to small welfare variations in this section. Reforms in market structure tend to benefit the poor more than the non-poor for maize, sorghum and wheat, the importable crops, whereas coffee and teff are neutral. A rise in the international price in the case of all importable crops under analysis also hurts producers, which means producers of these food crops do not in average, produce less than what they consume, buying the difference in the market. Finally, the effects tend to be small, since all of the crops have supply chains with a high degree of competition, which leads to small price variations when changing market structure, as we saw in the previous section.

Table 6.12 Sorghum price changes and household welfare

	Baseline	Leader split	Leaders merge	Exit of largest	Equal market shares	Perfect competition
Total						
Competition policy	0.00	0.00	0.00	0.00	0.00	0.05
International price	−0.34	−0.34	−0.34	−0.34	−0.34	−0.33
Poor						
Competition policy	0.00	0.00	0.00	0.00	0.00	0.06
International price	−0.46	−0.46	−0.46	−0.46	−0.46	−0.44
Non poor						
Competition policy	0.00	0.00	0.00	0.00	0.00	0.04
International price	−0.27	−0.27	−0.27	−0.27	−0.27	−0.26
Producers						
Competition policy	0.00	0.00	0.00	0.00	0.00	0.10
International price	−0.72	−0.72	−0.72	−0.72	−0.72	−0.69

Note: First order impact on household welfare

Table 6.13 Wheat price changes and household welfare

	Baseline	Leader split	Leaders, merge	Exit of largest	Equal market shares	Perfect competition
Total						
Competition policy	0.00	0.00	0.00	0.00	0.00	0.10
International price	−0.15	−0.15	−0.16	−0.16	−0.15	−0.07
Poor						
Competition policy	0.00	0.00	0.00	0.00	0.00	0.10
International price	−0.16	−0.16	−0.16	−0.16	−0.16	−0.07
Non poor						
Competition policy	0.00	0.00	0.00	0.00	0.00	0.09
International price	−0.15	−0.15	−0.15	−0.15	−0.15	−0.07
Producers						
Competition policy	0.00	0.00	0.00	0.00	0.00	0.09
International price	−0.15	−0.14	−0.15	−0.15	−0.15	−0.07

Note: First order impact on household welfare

References

Aguilar A, Carranza E, Goldstein M, Kilic T, Oseni G (2015) Decomposition of gender differentials in agricultural productivity in Ethiopia. Agric Econ 46(3):311–334

Bezabih M, Köhlin G, Mannberg A (2011) Trust, tenure insecurity and land certification in rural Ethiopia. J Socio-Econ 40:833–843

Clarke D, Coll-Black S, Cooney N, Edwards A (2016) A methodology to assess indicative costs of risk financing strategies for scaling up Ethiopia's productive safety net program. World Bank Policy Research Working Paper No. 7719

CSA (Central Statistical Agency) (2012) Agricultural sample survey 2011/2012: report on area and production of major crops. Central Statistical Agency, Addis Ababa

CSA (Central Statistical Agency) (2013) Agricultural sample survey: area and production of major crops, Meher season, vol I. CSA, Addis Ababa

Deininger K, Ali A, Alemu T (2011) Impacts of land certification on tenure security, investment, and land market participation: evidence from Ethiopia. Land Econ 87(2):312–334

Demeke M, F. Di Marcantonio (2013). Analysis of incentives and disincentives for sorghum in Ethiopia. Technical notes series, MAFAP, FAO, Rome

Dorosh P, Radish S (2012) Introduction, food and agriculture in Ethiopia, progress and policy changes. In: Dorosh P, Radish S (eds) IFPRI, published by University of Pennsylvania Press, Philadelphia, PA

FSCD (Food Security Coordination Directorate) (2007) Productive safety net program: logframe. Ministry of Agriculture and Rural Development, Federal Democratic Republic of Ethiopia, Addis Ababa

Headey D, Dereje M, Taffesse A (2014) Land constraints and agricultural intensification in Ethiopia: a village-level analysis of high-potential areas. Food Policy 48:129–141

ICO (International Coffee Organization) (2014) Annual review 2013–2014: strengthening the global coffee sector through international cooperation

Minten B, Dereje M, Engeda E, Kuma T (2015) Coffee value chains on the move: evidence from smallholder coffee farmers in Ethiopia (No. 76). International Food Policy Research Institute (IFPRI)

NBE (National Bank of Ethiopia) (2014) Annual report 2013–2014. NBE, Addis Ababa

Rashid S, Getnet K, Lemma S (2010) Maize value chain potential in Ethiopia: constraints and opportunities for enhancing the system. IFPRI Working Paper, Washington, DC

Shiferaw B, Negassa A, Koo J, Wood J, Sonder K, Braun J-A, Payne T (2011) Future of wheat production in Sub-Saharan Africa: analyses of the expanding gap between supply and demand and economic profitability of domestic production. In: Paper presented at the agricultural productivity-Africa conference, 1–3 November, 2011, Africa Hall, UNECA, Addis Ababa, Ethiopia

Taffesse A, Dorosh P, Gemessa S (2012) Crop production in Ethiopia: regional patterns and trends. In: Dorosh P, Radish S (eds) Food and agriculture in Ethiopia, progress and policy changes. IFPRI, published by University of Pennsylvania Press, Philadelphia, PA, pp 53–83

Vandercasteelen J, Dereje M, Minten B, Taffesse A (2013) Scaling-up adoption of improved technologies: the impact of the promotion of row planting on farmers' teff yields in Ethiopia

World Development Indicators (2016) World Bank, Washington, DC

Chapter 7
The Case of Uganda

7.1 The Household Survey Data

The household data comes from the Uganda National Household Survey 2005/ 2006. The dataset contains over 43,000 observations corresponding to more than 7,000 households. The overwhelming majority of the observations (78.48 %) and inhabitants in Uganda (84.10 %) are rural. Uganda's population is young, even by African standards: 50 % of the population is 14 or below and only 2.9 % is 65 or above. There are also slightly more females (51.4 %) than males, but not when it comes to household heads (26.9 %). Household sizes are large, also by African standards, 7.59 people per household, rural households being slightly larger than urban ones (7.64 versus 7.34).

Figure 7.1 shows the distribution of income. The graph shows the estimated density function of the logarithm of household per capita expenditure at the national level and for urban and rural regions separately. As it usually happens, the density for urban areas lies to the right of the density for rural areas, thus indicating that urban households have, on average, a higher level of expenditure per capita than rural households. Since the rural sample is bigger, the national distribution of income lies close to the rural density. The difference in the distributions is significant.

We now turn to sources of income and patterns of consumption. We start by presenting Table 7.1, which reports consumption patterns for both urban and rural regions. The usual regularities hold: auto consumption is much larger in rural (33.2 % of income) than in urban households (6.5 %). So is food consumption (63.5 versus 48.1 %). The most relevant crops when it comes to budget shares are Cassava (6.3 % of total expenditure), Yam (6.2 %), Maize (6.0 %), Livestock (3.6 %), Poultry (2.8 %), Rice (1.1 %), Millet (1.1 %) and Sorghum (0.8 %).

In Table 7.2 we present the same information, but from the income side. A similar pattern emerges. Auto-consumption remains more important for rural

N. Depetris Chauvin et al., *Agricultural Supply Chains, Growth and Poverty in Sub-Saharan Africa*, Advances in African Economic, Social and Political Development, DOI 10.1007/978-3-662-53858-6_7

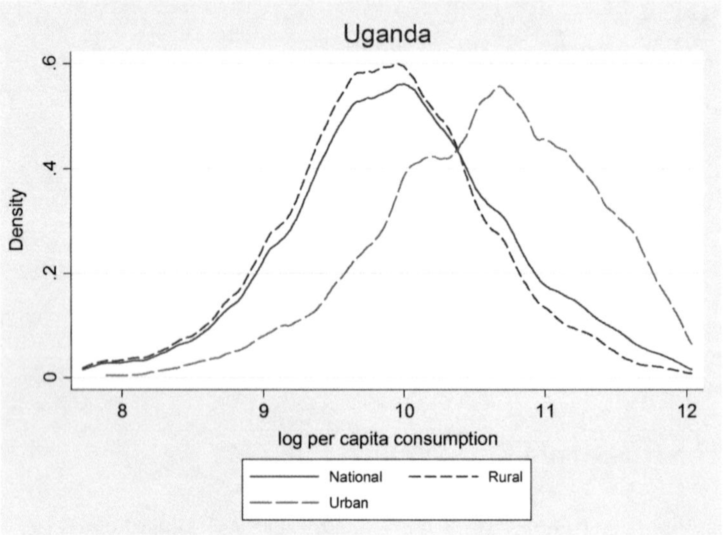

Fig. 7.1 The distribution of income density of (log) per capita household expenditure. Source: Uganda National Household Survey 2005/2006

Table 7.1 Budget shares

Uganda	Total	Rural	Urban
Total consumption per capita	100.0	100.0	100.0
Expenditures	70.9	66.8	93.5
Food	32.6	30.9	41.7
Manufactures	15.7	14.4	22.9
Services	22.6	21.5	28.9
Others	0.0	0.0	0.0
Auto-consumption	29.1	33.2	6.5
Auto-consumption food	28.6	32.6	6.3
Auto-consumption others	0.5	0.6	0.2
Total food consumption	61.2	63.5	48.1
Total crops	28.3	30.2	18.0
Cassava	6.3	7.0	2.6
Coffee	0.0	0.0	0.0
Cowpea	0.3	0.3	0.2
Livestock	3.6	3.7	3.3
Maize	6.0	6.3	4.2
Millet	1.1	1.3	0.3
Poultry	2.8	2.9	2.5
Rice	1.1	1.0	2.0
Sorghum	0.8	0.9	0.4
Wheat	0.0	0.0	0.0
Yam	6.2	6.9	2.5

Source: Uganda National Household Survey 2005/2006

households (29.6 versus 5.9 %) and the difference in food income is larger: (47.1 versus 11.1 %). The relative relevance of the crops also changes: Maize ranks first (5.5 % of total income), followed by Yam (5.1 %), Cassava (4.8 %), Coffee (2.6 %), Livestock (1.6 %), Poultry (1.5 %), Millet (1.1 %), Rice (0.7 %) and Sorghum (0.5 %).

Since we are not only interested on the overall effects of changes in prices but also on the impact on households with different levels of income, we provide information on these shares for different income levels, using the log of per capita consumption (log pce) as our measure of well-being. We plot estimates of non-parametric regressions of income and budget shares on log pce.

We start with a broad measure: the average of food consumption (expenditure plus auto-consumption) and the average of food income (again cash income and auto-consumption) in Figs. 7.2 and 7.3 respectively. We see that food expenditure is declining in consumption, and for the most part, rural expenditure is higher than urban expenditure shares. At the top of the income distribution they converge to approximately 30 %, from a maximum of 65 % for urban households and 70 % for rural ones. Food income is different. For both type of households, it has a definite inverted u shape. Rural households derive a significant larger part of their income from food production, totaling almost 60 % at the middle of the income distribution, whereas urban household's maximum is 20 %.

Table 7.2 Income shares

Uganda	Total	Rural	Urban
Total income per capita	100.0	100.0	100.0
Incomes	74.0	70.4	94.1
Food (agriculture)	16.1	18.1	5.4
Wage	19.1	16.4	33.6
Enterprises	19.4	16.3	36.1
Transfers	19.5	19.6	19.1
Auto-consumption	26.0	29.6	5.9
Auto-consumption food	25.5	29.1	5.7
Auto-consumption others	0.5	0.6	0.2
Total food income and AC	41.6	47.1	11.1
Total crops	23.6	26.8	6.4
Cassava	4.8	5.5	1.1
Coffee	2.6	3.0	0.6
Cowpea	0.1	0.2	0.0
Livestock	1.6	1.8	0.4
Maize	5.5	6.1	1.8
Millet	1.1	1.3	0.2
Poultry	1.5	1.7	0.4
Rice	0.7	0.8	0.2
Sorghum	0.5	0.6	0.1
Wheat	0.0	0.0	0.0
Yam	5.1	5.7	1.5

Source: Uganda National Household Survey 2005/2006

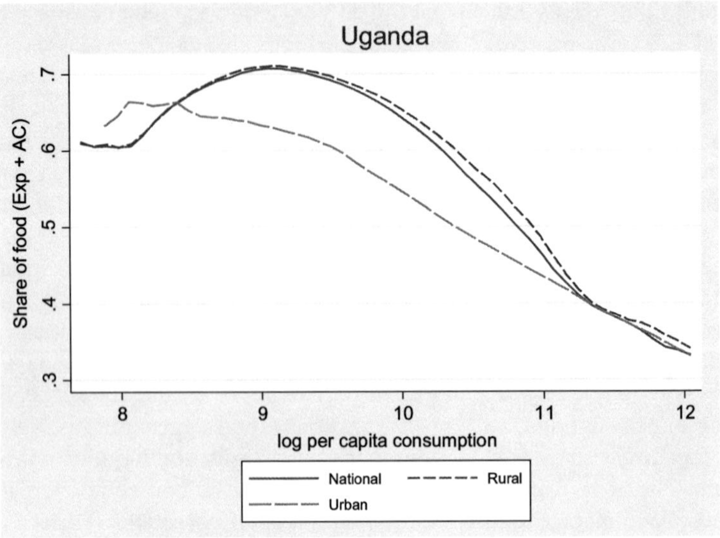

Fig. 7.2 Total food budget share across the income distribution. Source: Uganda National Household Survey 2005/2006

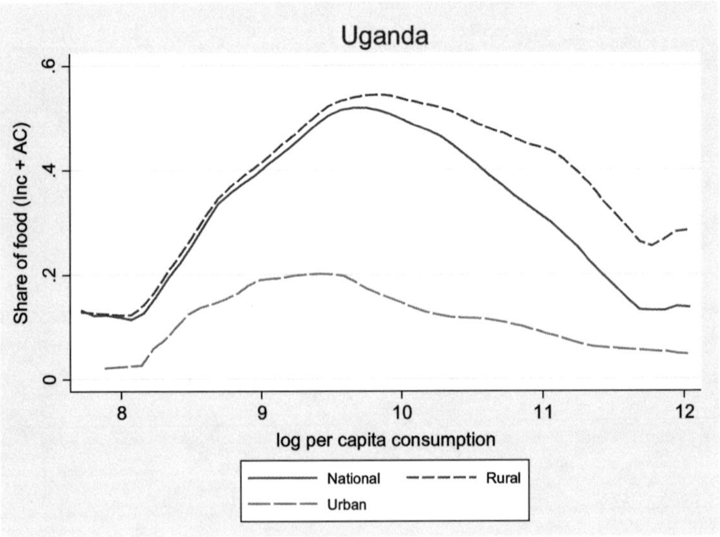

Fig. 7.3 Total food income share across the income distribution. Source: Uganda National Household Survey 2005/2006

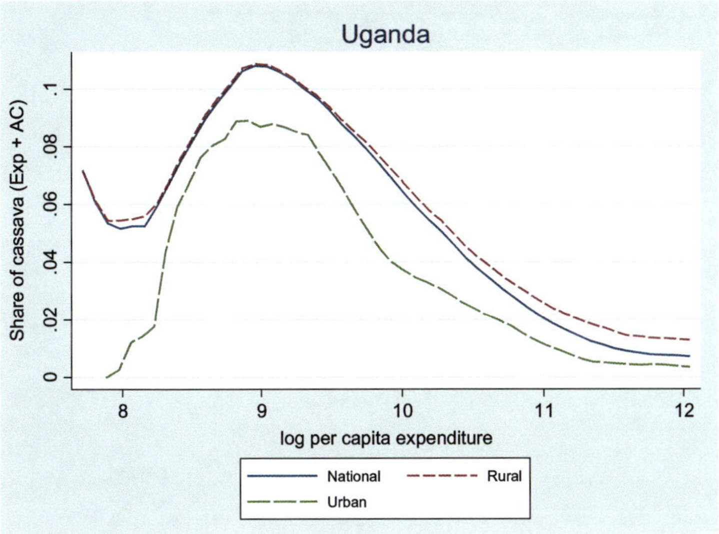

Fig. 7.4 Cassava budget share across the income distribution. Source: Uganda National Household Survey 2005/2006

We focus now on the target products in our simulations (Figs. 7.4, 7.5, 7.6, 7.7, 7.8, 7.9, 7.10, 7.11, 7.12, 7.13 and 7.14). We start with the food crops. Cassava's budget shares (Fig. 7.4) and income shares (Fig. 7.5) follow an inverted u pattern, for both type of households. Budget shares, however, are at a higher level than income shares. Urban shares are below rural ones in both cases.

Livestock's budget and income shares (Figs. 7.6 and 7.7) are increasing in income for both types of households. However, in the case of budget shares, urban and rural ones are at the same level, whereas rural income shares are significantly higher than urban shares. As it was with cassava, budget shares are higher than income shares.

In Figs. 7.8 and 7.9 we show the budget and income shares for maize. The budget share for urban households is decreasing in income, while the budget share for rural households follows an inverted u shape. Urban shares start at a higher level than rural ones, to then converge to rural levels. Rural households show the same pattern in their income as in their expenditure, an inverted u shape, although the maximum is reached at a higher level of income. But urban households have a slightly decreasing pattern, below the level of the rural households for most of the distribution.

Millet's budget (Fig. 7.10) and income (Fig. 7.11) shares are almost identical. They both follow an inverted u shape, for rural and urban households, with urban households' shares at a lower level than rural ones. Budget shares are above income shares, however, around 2 % versus 1.5 %, respectively.

We show sorghum's budget share and income shares in Figs. 7.12 and 7.13. The budget shares are decreasing in income for rural and urban households, but they converge to zero in both cases at the middle of the income distribution. Income shares for rural households have a similar pattern, although for urban ones there is an inverted u shape, with most households having a share close to zero.

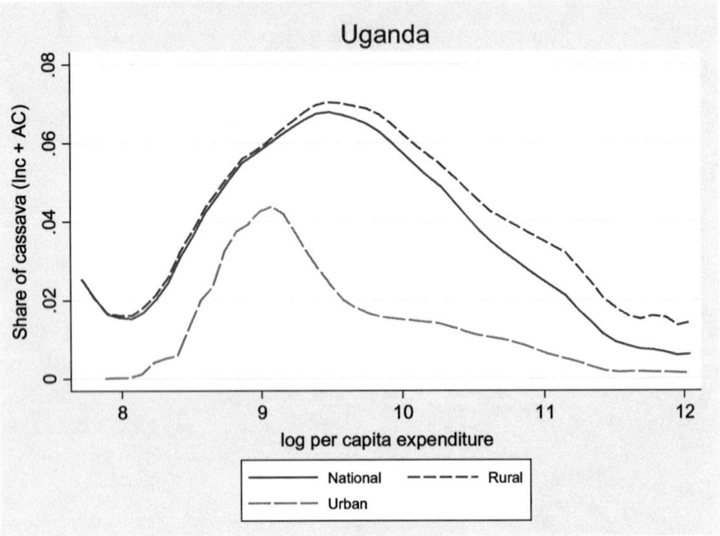

Fig. 7.5 Cassava income share across the income distribution. Source: Uganda National Household Survey 2005/2006

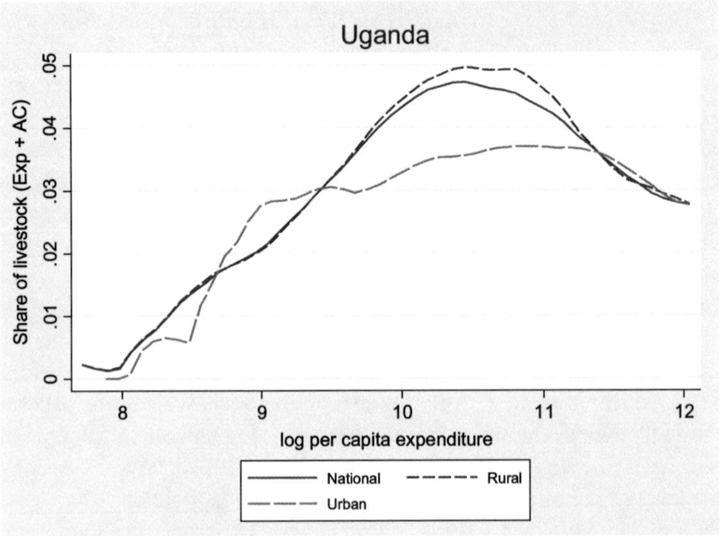

Fig. 7.6 Livestock budget share across the income distribution. Source: Uganda National Household Survey 2005/2006

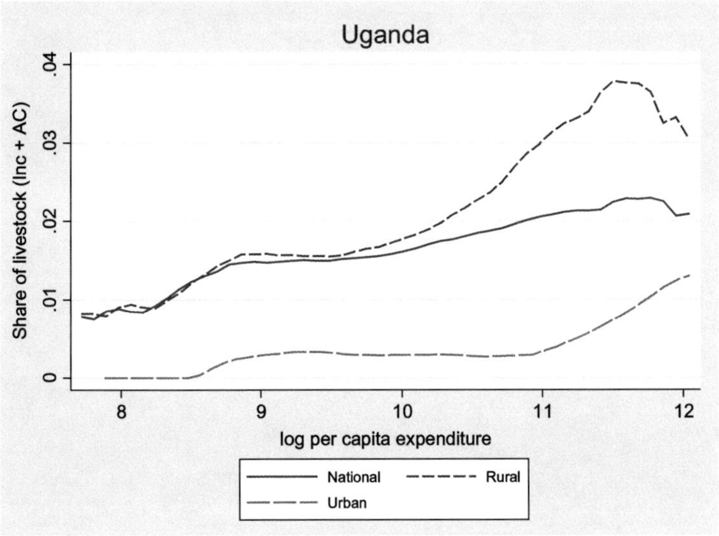

Fig. 7.7 Livestock income share across the income distribution. Source: Uganda National Household Survey 2005/2006

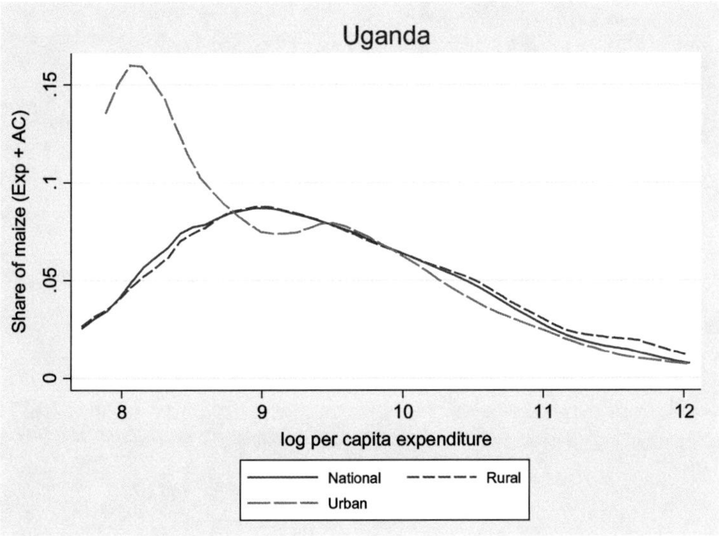

Fig. 7.8 Maize budget share across the income distribution. Source: Uganda National Household Survey 2005/2006

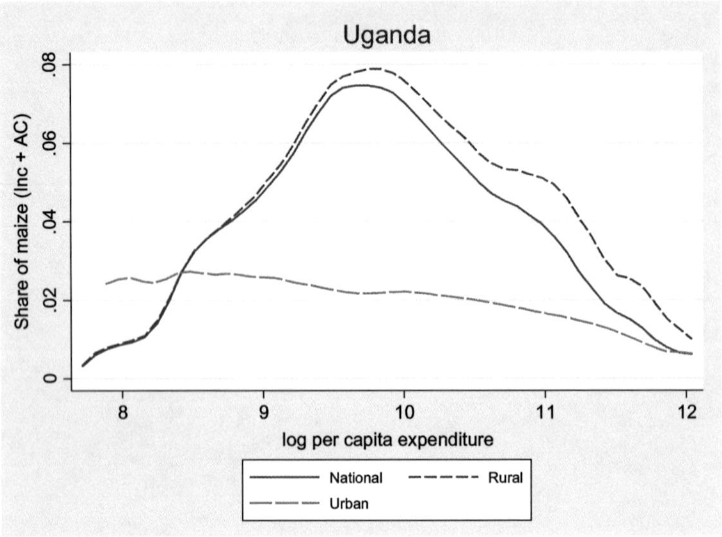

Fig. 7.9 Maize income share across the income distribution. Source: Uganda National Household Survey 2005/2006

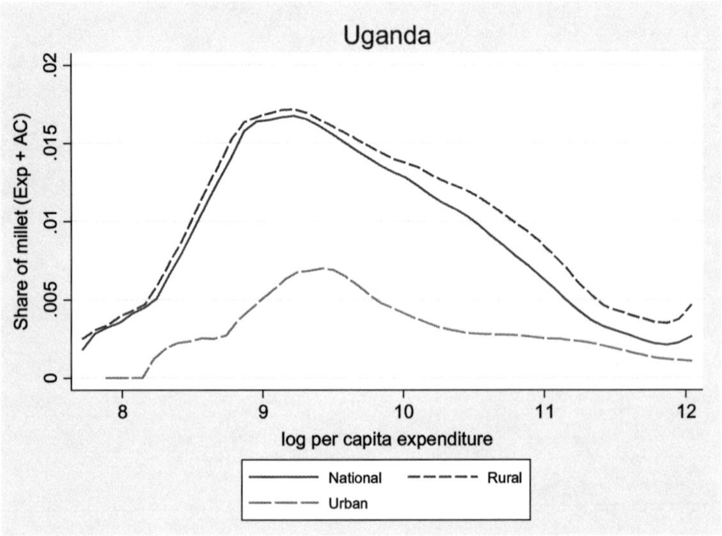

Fig. 7.10 Millet budget share across the income distribution. Source: Uganda National Household Survey 2005/2006

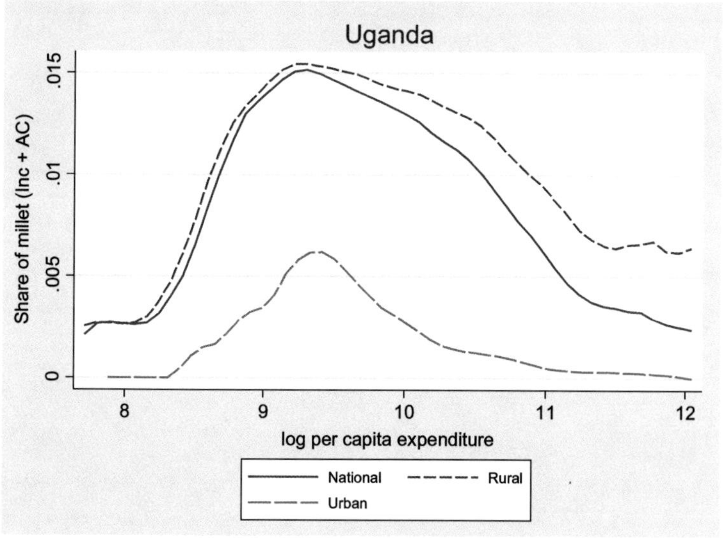

Fig. 7.11 Millet income share across the income distribution. Source: Uganda National Household Survey 2005/2006

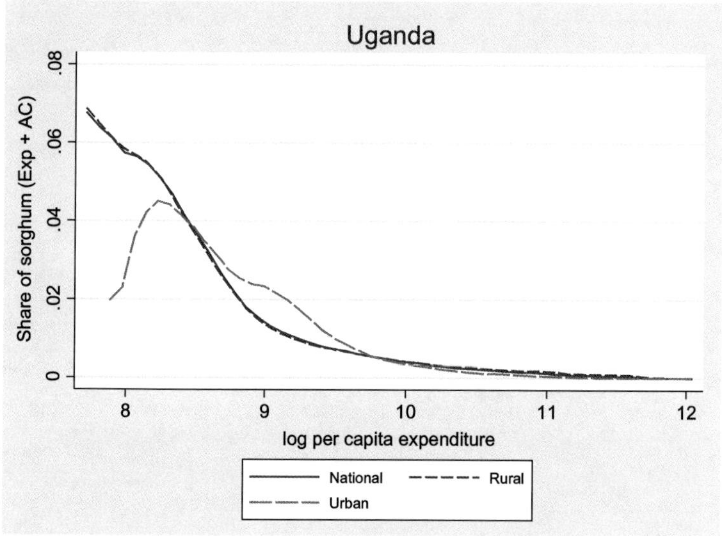

Fig. 7.12 Sorghum budget share across the income distribution. Source: Uganda National Household Survey 2005/2006

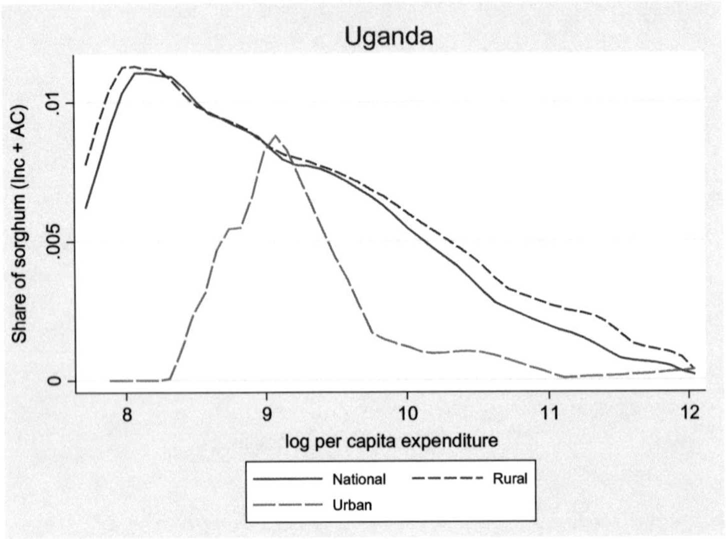

Fig. 7.13 Sorghum income share across the income distribution. Source: Uganda National Household Survey 2005/2006

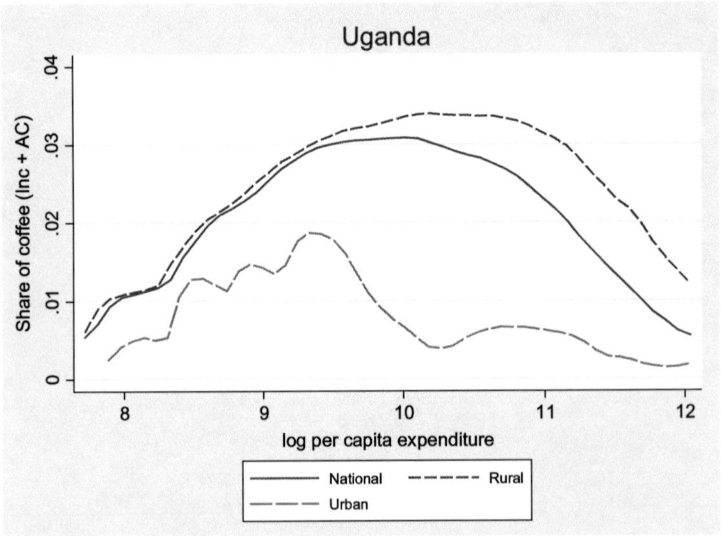

Fig. 7.14 Coffee income share across the income distribution. Source: Uganda National Household Survey 2005/2006

Finally, we present the income share of our only cash crop for Uganda, coffee, in Fig. 7.14. Rural shares have an inverted u shape, reaching 4 % of income at the middle of the income distribution. Urban shares are irregular because few households grow coffee.

7.2 Food Crops, Cash Crops, and Livestock in Uganda

Agriculture is a core sector of Uganda's economy. It contributed 25.9 % of GDP at current prices in 2012 and exports from the sector accounted for 48.5 % of total exports in 2012. About 60 % of Uganda's population is engaged in agriculture. The agricultural sector is fragmented and dominated by small farmers most of whom combine subsistence farming with cash crop and livestock farming. Traditional cash crops in Uganda include coffee, cotton, tea, cocoa, tobacco and sugarcane. Non-traditional cash crops like maize, rice, beans, soya beans, palms, and horticultural produce are also important. Livestock, in particular cattle and goats, is also important (Figs. 7.15 and 7.16).

For the last 25 years, a number of policies, strategies, and programs have been put in place to support the food and agriculture sector in Uganda. MAFAP (2013) categorized them into overarching national policies and specific agricultural and rural development initiatives. Among the specific agricultural and rural development initiatives, the Plan for the Modernization of Agriculture (PMA) stands out to be key. It was designed as a strategic framework for eradicating poverty through the pursuance of multi-sectoral interventions to tackle a wide range of multi-faceted constraints that farmers face. The PMA is part of the Government of Uganda's strategy of poverty eradication as contained in the Poverty Eradication Action Plan. The main objectives of the PMA, as outlined by Kraybill and Kidoido (2009), are: (1) To increase incomes and improve the quality of life of poor subsistence farmers through increased productivity and increased marketing of farm production; (2) To improve household food security through the market rather than emphasizing self-sufficiency; (3) To provide gainful employment through agro-processing factories and services, and; (4) To promote sustainable use and management of natural resources by developing improved land use and land management policy and by promoting environmentally friendly technologies.

Among the crops recommended to be targeted in the PMA cassava is mentioned. Cassava is one of the most important sources of energy in the diet of many people living in the tropics. In Uganda cassava plays a very important role in both household and national food security. It is grown in every part of the country and used mainly as food and in a variety of ways such as local brews, feed, plywood industries and as high quality cassava flour (HQCF). It is emerging as a raw material in bakeries, breweries, paperboard and composite flours. The ability of cassava to grow well in marginal environment (particularly in drought prone areas), its apparent resistance to pests and diseases, flexibility in the cropping and food systems makes it an attractive crop to most Ugandan farmers. Nationally, the crop generates over 60 % of the basic food requirements for the North, Eastern

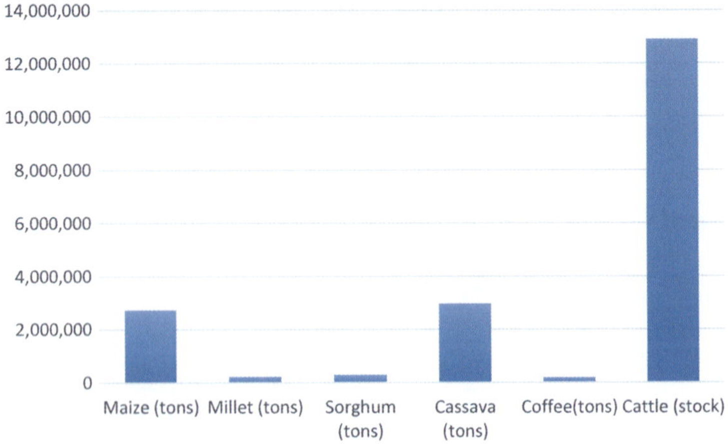

Fig. 7.15 Crops and livestock production in Uganda (in tons, 2013). Source: FAOSTAT online

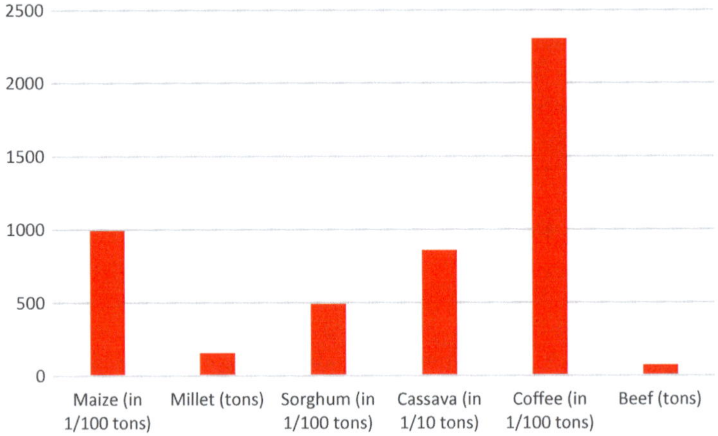

Fig. 7.16 Crops and livestock net export in Uganda (in tons, 2013). Source: FAOSTAT online

and West Nile regions of the country. Cassava is the second most important staple crop after bananas and Uganda is the sixth largest producer of the crop in Africa, with 4.2 million metric tons in 2010 and a decline to 2.9 million in 2013. Most of the production is consumed domestically with exports only been residual (net exports were around 8000 t in 2013).

Two of the major constraints to cassava value chain are (1) the perishability of the fresh roots and (2) the presence of cyanogenic compounds in cassava. Cassava roots that are more than 48 hours old have little market value and limited marketability. Therefore, communities faced with limited access to processing units would not be able to effectively commercialize cassava. Fresh cassava contains cyanogenic glucosides and if cassava is inadequately processed this creates a potential health hazard. Effective processing, essentially involving root disintegration and removal of the cyanogenic compounds with the water, ensures the safety of products (Nhassico et al. 2008). However, availability of cassava processing units as mentioned earlier is very limited.

The livestock sector in Uganda contributes about 9 % to total agricultural output and 2 % to total gross domestic product. Livestock, more especially cattle, play multiple roles and provide many valuable services and products for rural households as it is regarded as an asset, source of wealth and a major contributor to food security. Cattle is the main source of animal protein (milk and meat) in the country. In most communities in Uganda where agro-pastoralism is practiced, cattle dung is the main and preferred source of soil nutrient replenishment. In addition, in Eastern and Northern Uganda, cattle (bulls) are used for traction. According to the Uganda Census of Agriculture 2008/2009, up to 26 % of households in the country own cattle, 39 % own goats, 9 % own sheep and 18 % own pigs. Of all domestic animal meats, estimates indicate that cattle beef is the most consumed meat in the country followed by pork. The majority of cattle farmers are smallholders who rear cattle primarily for milk production, most of which is consumed at home. Cattle farmers only sell culled cattle to traders or butchers for beef in a one-off transaction and as such they are not integrated into the beef value-chain. In 2011 production of beef was 185,700 t and net exports only 34 t.

Farm-level livestock productivity is very low. It is mostly attributed to factors related to the lack of commercial orientation among pastoralists (ACET 2015). In fact, most farmers keep livestock for prestige and for milk, and so tend to have more cows than bulls, while in commercial oriented beef production, bull are preferred because of their weight. In addition, provision of public veterinary services is weak. This then leaves the duty to private sector actors whose services are not affordable to the average beef producer. Low quality pasture is also prevalent due to overgrazing and frequent drought conditions which have increased competition for water and pasture.

In Uganda, finger millet is the third most important cereal after maize and sorghum occupying more than 400,000 ha. It is the staple food for over 50 % of the country's population and increasingly a major source of income. It grows in all ecological areas of the country but the northern region is the leading growing area, accounting for 40 % of total production, followed by the eastern region with 20 % of the output. Millet grains store well and for long periods without the risks of insect damage due to the small size of its seed, which make it an ideal food security crop and an important preventive food against malnutrition. The country is an exporter of millet and its processed sub products but those processed products are largely confined to the African supermarkets because they lack certification and quality marks.

Millet yields are generally below potential. Some of the reasons behind it includes the declining soil fertility in millet producing districts. While this could be remedied through soil enriching practices such as conservation agriculture, the situation is made worse because millet is not among the targeted crops in agricultural policies, including the PMA discussed earlier.

Sorghum is the last food crop we analyze for Uganda. The crop is uniquely adapted to Africa's climate, being both drought resistant and able to withstand periods of water-logging. In Uganda, the crop is the third most important after maize and cassava. Annual production in 2012 was estimated at 315,000 t from about 350,000 ha. It is important both for food security and for cash. In Uganda, sorghum is used for seeds, human consumption, animal feeds and industrial purposes. A variety of products are made from sorghum including composite flours, local beverages and commercial beer. The figures for sorghum trade are not well established as a large percentage of the trade in the commodity is not recorded and for this reason exact quantities exported and imported regionally are not well established. According to FAO, the country was a net importer of sorghum in 2011 but anecdotal evidence suggests the country is actually a net exporter.

Among the challenges faced by sorghum producers we note the prevalence of fake seeds. ACET (2015) noted that about 40 % of seed marketed in Uganda are fake. As a result, farmers show poor adoption of improved seeds, even though they may be authentic, because of loss in faith of the seed multiplication system. Bird menace is also a challenge in the sorghum production, especially to the white seeded sorghum mostly grown by commercial farmers. It has been reported that in 2013 birds were responsible for nearly 60 % of sorghum production loss (ACET 2015).

Coffee is Uganda's principal export and accounts for over 25 % (on average) of the country's total export earnings. The total production in 2014 stood at 190,000 t. The Uganda Census of Agriculture (UCA) estimates that over 98 % of the coffee farmers in Uganda grow coffee in areas of 0.2 ha or less. About 1.8 % of the coffee farms have land size of up to 15 ha, and the rest, about 0.2 % of the total national production is grown in a large farm in Mubende District Mid Western Uganda called Kaweri farms. According to the Uganda Investment Authority, Robusta coffee is grown in the low altitude areas of Central, Eastern, Western and South Eastern Uganda up to 1200 m while Arabica coffee is grown in the highland areas on the slopes of Mount Elgon in the East and Mt. Rwenzori and Mt. Muhabura in South Western Region Uganda at an altitude of 1500–2300 m. Arabica and Robusta coffee on average contribute 15 % and 85 % of annual coffee production respectively. The coffee industry employs over 3.5 million families who live and work on these farms and in related support and downstream businesses. Robusta yields in Uganda are about 648 kg/ha. In comparison, average Robusta yields in Vietnam are 2.2 t/ha. Arabica does not fare any better at about 800 kg/ha compared to over 2000 kg/ha in Brazil.

According to USAID (2011), three major constraints that the Uganda coffee value chain has faced over the past years include (1) coffee wilt disease (CWD) in Robusta, (2) low coffee prices of 2001 and (3) soil fertility depletion in coffee producing areas. CWD is associated with the decimation of half of Uganda's Robusta trees in 2001. While local research institutions have found the cure against this, the main challenges that remains is the multiplication of the developed pest resistant plants. This could be due to the low coffee prices that have led many farmers to diversify away from coffee leading to a gradual reduction in coffee

Table 7.3 Market shares in Uganda

Cassava		Beef		Coffee	
Company	Shares (%)	Company	Shares (%)	Company	Shares (%)
Local millers	96.60	Fresh Cuts	86.00	Ugacof Ltd	15.01
NUMA Feeds	0.80	Sausage King	7.00	Olam	12.00
Maganjo	0.60	Your Choice	7.00	Pan Afric	8.69
Family Diet	0.30			IberoLtd	8.00
Pallisa Agri-business Training Association	0.20			LD Commodities	7.98
MACO ltd	0.20			Kyagalanyi Coffee	7.91
Others	1.30			Kampala Domestic Stores	5.96
				Kawacom Ltd	5.76
				Savannah Commodities	4.34
				Lakeland Holdingd	4.22
				Others	20.12

Millet		Sorghum	
Company	Shares (%)	Company	Shares (%)
Local millers	48.30	Local millers	49.40
Other processing companies	27.60	Uganda Breweries	17.10
Unga Kenya Ltd millers	14.50	Other pocessing companies	14.40
Maganjo	3.60	Niles Breweries	14.20
East Africa Basic Foods	1.20	NUMA Feeds	1.70
NUMA Feeds	1.20	Katumba	1.00
Katumba	0.60	Ndungu	0.90
Family Diet	0.50	Mzungu	0.90
MACO Ltd	0.50	3R Agro industries ltd	0.10
Majid	0.50	HARREE Millers U ltd	0.10
SESACO Ltd	0.20	Opit Investment millers Lts	0.10
EDRAC Ltd	0.20		
Ndungu	0.20		

Source: Stakeholders interviews

production. Also, the low soil fertility in coffee producing areas could also have exacerbated coffee production declines.

Table 7.3 below shows the market shares of processors for the crops under study in Uganda. In livestock, the processors are concentrated. Fresh cuts produces 86 % of the total processed meat followed by Sausage King and Your Choice that have

7 % each of the market. For the other agricultural products processors are far less concentrated. Modern processing consists of mills that turn millet grain, sorghum grain and cassava chips into flour. There are three types of processors. The first are service millers who process for others at a fee. Most millers utilize hammer mill technology, are small scale (<10 t/month capacity) and constitute 85 % of millers in the country and handle over 50 % of the total produce. These are not included in the list. The second type are the trade-based millers who in addition to providing the milling service also purchase either milled flour, or clean grain or cassava chips from the traders, which they then use as an input in their composite flours. The last category of processors develop millet, sorghum and cassava based products and package for middle and high income earners in urban and peri-urban areas as well as for export. Although most of the products are composite flours, some of the new millers have ventured into alcoholic and non-alcoholic beverages. Of the three crops, sorghum is the one there is more market concentration. While almost 50 % is processed by local millers, three companies account for 46 % of the market. In millet, Unga Kenya Ltd is the largest company processor with 14.5 % of the market, followed by Maganjo with 3.6 % of the market. In cassava there is no a dominant company at the national level as local millers account for almost all the processing. For coffee, since liberalization of the market in 1992, farmers have been free to decide how and to whom to sell their coffee. For the majority of farmers the price is negotiated at the time of sale and payment is not made until then. Transactions at the farm level are quite small and farmers usually sell as individuals, only in a few cases selling as a group. Most coffee sales are made at the farm gate to small traders. These small-scale traders act as aggregators either for bigger independent traders, who often own a store or mill, or for exporters and their agents (Vargas Hill 2010). After the coffee has been milled, it is transported to Kampala and sold to exporters. At the export level, around 75 % of the volume was handled by ten companies, with the largest being Ugacof Ltd. (15.01 %), Olaf (12 %), and Pan Afric (8.69 %). Competition among coffee exporters in Uganda has steadily increased in the last few years.

Farm gate prices for the products in this study are dictated by the buyers on-farm and local markets. Usually at harvest when there is a glut buyers offer very low prices as much as half of the highest when the commodity is scarce. The harvest periods usually coincide with back-to-school and the farmers are desperate to sell the crops to get school fees. However when the crop is scarce the farmers can bargain and get good prices especially if they were able to store it or process it into a form that gives it a longer shelf-life. Government of Uganda liberalization the marketing sector and does not have any role in pricing and storage facilities. Hence there are no government buffer stocks.

Table 7.4 Farmgate price simulation results for coffee

	Baseline	Leader split	Leaders merge	Exit of largest	Equal market shares	Perfect competition
Competition policy	0.00	0.19	−0.19	−0.23	1.55	3.14
Increase of 10 % in						
International price	15.54	15.78	15.31	15.27	17.13	19.27
Marginal cost of producing cash crop	0.22	0.41	0.07	0.03	1.77	3.14
Fixed cost of producing cash crop	0.12	0.31	−0.06	−0.10	1.68	3.14
Household resources (endowment)	−0.29	−0.07	−0.49	−0.53	1.29	3.14
Risk and food security parameter	0.00	0.20	−0.18	−0.22	1.56	3.14
Food crop price	0.21	0.38	0.03	−0.01	1.76	3.14
Marginal cost of producing food crop	−0.03	0.17	−0.21	−0.26	1.53	3.14
Non-farmer demand	0.00	0.19	−0.19	−0.23	1.55	3.14

Source: Simulation results from the model of Chap. 2

7.3 Simulation Results

In this section, we present the simulation results for the target crops. We classify these crops in cash crops (coffee), exportable food crops (cassava, livestock, maize, millet) and importable food crops (sorghum).

We start with coffee. Coffee's supply chain is relatively competitive, which is unusual in cash crops. The largest firm has a market share of 15 %. Therefore, farmgate price change results should be smaller than in previous cash crop simulations. This is indeed the case, as we show in Table 7.4. In row 1, where we present the changes in market structure, the largest result, a change towards perfect competition in the supply chain, leads only to a 3.14 % increase in farm gate prices. A rise in international prices of 10 %, shown in row 2, leads to an increase of 15.54 % in farm gate prices. The rest of the effects following shocks to domestic constraints are quite limited, none above 1 % in absolute value. As we have discussed in previous countries, the sum of two shocks does not necessarily equal the sum of the effects of the shocks, complementarities and substitutabilities arise. An example of the former is a Rise of 10 % in international prices paired with convergence towards perfect competition: the combined effect is 19.27 % but the sum of the effect is 18.68 %. An example of the latter is the combination of Perfect competition again and a rise in the marginal cost of producing the cash crop (combined effect 3.14 %, sum 3.36 %).

We move on to the exportable food crops. The first in our list is cassava, its results are in Table 7.5. Cassava's supply chain is not concentrated, therefore we should expect small results of changes in market structure as it is indeed the case for

Table 7.5 Farmgate price simulation results for cassava

	Baseline	Leader split	Leaders merge	Exit of largest	Equal market shares	Perfect competition
Competition policy	0.00	0.05	−0.05	−0.05	0.00	1.20
Increase of 10 % in						
International price	17.59	17.70	17.44	17.44	17.59	20.43
Marginal cost of producing cash crop	−0.11	−0.05	−0.17	−0.17	−0.11	1.20
Fixed cost of producing cash crop	−0.47	−0.40	−0.56	−0.56	−0.47	1.20
Household resources (endowment)	0.41	0.44	0.38	0.38	0.41	1.20
Risk and food security parameter	0.42	0.46	0.39	0.39	0.42	1.20
Cash crop price	1.06	1.07	1.06	1.06	1.06	1.20
Marginal cost of producing food crop	0.33	0.37	0.29	0.29	0.33	1.20
Non-farmer demand	0.02	0.07	−0.03	−0.03	0.02	1.20

Source: Simulation results from the model of Chap. 2

all results in Row 1. Row 2, where we report the effects of an increase in international prices, shows that an increase of international prices of 10 % leads to an increase of farm gate prices of 18.75 %. Results of other parameters in Rows 3–9 are small, as usual. An example of a complementarity is the combination of Perfect Competition with International prices (combined 20.43 %, sum 18.78 %) and for substitutability an increase in Perfect Competition, again with an increase in the price of the Cash Crop (combined 1.20 %, sum 2.26 %).

The second exportable food "crop" we analyze is livestock. The results for the simulation are shown in Table 7.6. Livestock's supply chain is the one of the most concentrated chains in the country, the largest firm having a market share of 86 %, and the second largest 7 %. Therefore, the price change results presented in the simulation are large. In row 1, for example, changes that were usually small in the simulations we have been running are very relevant, as Exit of Largest (−7.09 % variation in farm gate prices), or Equal Market Shares (10.48 %). The most extreme variation, Perfect Competition, leads to an impressive increase of 30.71 % in farm gate prices. Row 2 shows an elasticity of local prices with respect to international prices of 9.25 %. As we discussed before, low concentration means convergence to a model of a traded good, in which internal shocks in demand or supply have small effects on prices. Since livestock is at the other end of the spectrum (almost a monopoly) the internal shocks presented in Rows 3–9 do become much larger. An increase of a 10 % in the price of the Cash Crop, the rival good, leads to a 5.24 % increase in the farm gate price of livestock, a 0.524 cross elasticity. Complementarities and substitutabilities are also present as before, but the absolute difference between the joint effects and the sum of the effects are larger, because the effects are larger. The combination is of Perfect Competition with a rise in International

Table 7.6 Farmgate price simulation results for livestock

	Baseline	Leader split	Leaders merge	Exit of largest	Equal market shares	Perfect competition
Competition policy	0.00	4.83	−0.58	−7.09	10.48	30.71
Increase of 10 % in						
International price	9.25	15.01	7.16	0.76	19.17	47.38
Marginal cost of producing cash crop	−0.45	4.41	−1.26	−7.72	9.90	30.71
Fixed cost of producing cash crop	−1.94	3.22	−3.07	−9.56	8.73	30.71
Household resources (endowment)	1.26	5.74	1.04	−5.22	11.26	30.71
Risk and food security parameter	2.05	6.51	2.00	−4.22	12.26	30.71
Cash crop price	5.24	9.14	5.38	−0.86	15.91	30.71
Marginal cost of producing food crop	1.60	6.21	1.38	−5.34	12.39	30.71
Non-farmer demand	0.26	5.02	−0.28	−6.74	10.71	30.71

Source: Simulation results from the model of Chap. 2

Prices of 10 % lead to an increase of 47.38 % in local prices, whereas the sum is 39.96 %, a complementarity. A substitutability would be a rise of the Price of the Cash Crop and Perfect Competition, being the joint effect 30.71 % and the sum 35.95 %.

The third exportable food crop we include is maize, which we show in Table 7.7. Maize's supply chain is also very concentrated, the largest firm having a share of 52 % and the second largest 22 %. Therefore, a convergence to Perfect Competition leads to an increase of 13.15 % in farmgate prices, as we show in Row 1 with the rest of the changes in market structure. The elasticity of farm gate prices to international prices is slightly above 1, since a 10 % increase in international prices leads to a 10.11 % rise in local prices, as we show in Row 2. Due to the high concentration, some of the shocks presented in Rows 3–9 become significantly. Notably, everything related to the rival market, the cash crop, matters. An increase in marginal costs in the cash crop push the price of maize down by 4.64 %, whereas an increase in the price of the cash crop increases the price of maize by 6.51 %, a crossed elasticity even higher than with livestock. Complementarities and substitutabilities arise as always. An example of the former is a combination of a rise in international prices and Perfect Competition, the joint effect is 34.37 % and the sum only 23.26 %, a difference larger than 10 % points, which stresses the importance of taking complementarities into account. An example of substitutabilities is an increase in the price of the Cash Crop and Perfect Competition (combination 13.10 %, sum 19.66 %), a difference in more than 6 % points.

The last exportable food crop is millet. We present the results of the simulations for this crop in Table 7.8. Millet is not as concentrated as the last two crops we just

Table 7.7 Farmgate price simulation results for maize

	Baseline	Leader split	Leaders merge	Exit of largest	Equal market shares	Perfect competition
Competition policy	0.00	1.63	−0.96	−2.30	3.73	13.15
Increase of 10 % in						
International price	10.11	12.30	8.12	7.20	12.75	34.37
Marginal cost of producing cash crop	−4.64	−2.74	−6.06	−7.29	−1.56	13.10
Fixed cost of producing cash crop	−1.43	0.37	−2.52	−3.90	2.15	13.10
Household resources (endowment)	0.89	2.40	0.12	−1.04	4.52	13.10
Risk and food security parameter	0.43	1.89	−0.45	−1.81	4.02	13.10
Cash crop price	6.51	7.44	6.72	5.21	10.68	13.10
Marginal cost of producing food crop	2.65	3.98	2.12	0.79	6.69	13.10
Non-farmer demand	0.10	1.64	−0.90	−2.27	3.74	13.10

Source: Simulation results from the model of Chap. 2

Table 7.8 Farmgate price simulation results for millet

	Baseline	Leader split	Leaders merge	Exit of largest	Equal market shares	Perfect competition
Competition policy	0.00	0.29	−0.10	−0.31	3.46	6.03
Increase of 10 % in						
International price	17.79	18.15	17.61	17.40	21.30	25.26
Marginal cost of producing cash crop	−2.10	−1.72	−2.32	−2.54	1.36	6.03
Fixed cost of producing cash crop	−1.06	−0.72	−1.23	−1.44	2.41	6.03
Household resources (endowment)	0.92	1.17	0.87	0.66	4.37	6.03
Risk and food security parameter	0.82	1.07	0.77	0.55	4.25	6.03
Cash crop price	3.63	3.75	3.72	3.51	7.05	6.03
Marginal cost of producing food crop	0.24	0.52	0.14	−0.06	3.70	6.03
Non-farmer demand	0.02	0.31	−0.08	−0.29	3.48	6.03

Source: Simulation results from the model of Chap. 2

analyzed, the largest firm having 15 % of market share and the second largest 5 %. Therefore, we show in Row 1 along the rest of the market changes, that changing to Perfect Competition would lead to a modest increase of 6.03 % in farm gate prices.

Table 7.9 Farmgate price simulation results for sorghum

	Baseline	Leader split	Leaders merge	Exit of largest	Equal market shares	Perfect competition
Competition policy	0.00	−0.23	0.22	0.26	−2.32	−4.41
Increase of 10 % in						
International price	8.05	7.89	8.20	8.24	5.70	4.85
Marginal cost of producing cash crop	−3.16	−3.23	−3.12	−3.09	−5.45	−4.41
Fixed cost of producing cash crop	−0.86	−1.04	−0.70	−0.67	−3.19	−4.41
Household resources (endowment)	0.97	0.69	1.25	1.28	−1.31	−4.41
Risk and food security parameter	1.44	1.15	1.74	1.77	−0.80	−4.41
Cash crop price	5.13	4.64	5.64	5.67	2.85	−4.41
Marginal cost of producing food crop	0.33	0.09	0.57	0.60	−1.92	−4.41
Non-farmer demand	0.05	−0.18	0.27	0.31	−2.28	−4.41

Source: Simulation results from the model of Chap. 2

The elasticity of farm gate prices with respect to international prices is, however, large, 1.8 implied by the results in Row 2. From the rest of the results in Rows 3–9, we stress only a Rise of the price of the Cash Crop of 10 %, which leads to an increase of 3 % in the price of millet, a sizeable 0.3 crossed-elasticity. Complementarities and substitutabilities arise, but are lower in absolute value than those in maize. The usual complementarity takes place, Perfect Competition and increase in International Price has a combined effect of increasing farm gate prices 25.26 % but the sum of the effects is 23.82 %. And as substitutability, the combination of Perfect Competition and an increase in the Cash Crop price, the joint effect being 6.03 % while the sum is 9.66 %.

Finally, we analyze an importable food crop, sorghum. Its results are shown in Table 7.9. The supply chains of this crop are mildly concentrated, the two largest firms being similar in market shares, 16 and 14 %. We see in Row 1 that this makes for a moderate decrease in prices when we assume that market power vanishes and the market structure is transformed to Perfect Competition, the fall in prices being 4.41 %. In Row 2 we see elasticities of local prices with respect to international prices of 0.8 approximately. From the rest of the results in Rows 3–9 we mention those in Rows 4 and 5, where we see that an increase of 10 % in the marginal cost of the Cash Crop leads to a fall of 3.16 % in local prices and that an increase in the Price of the Cash Crop leads to an increase in sorghum farm gate prices of 5.13 %. The largest complementarity in absolute values is that which arises from the combination of Perfect Competition and increase of the Price of the Cash Crop, the joint effect being −4.41 % and the sum 0.72 %. As substitutabilities we mention Perfect Competition and increase of the Marginal Cost of the Cash Crop (combined −4.41 %, sum −7.58 %).

7.4 Welfare Simulations

We end our analysis with a discussion of the welfare and poverty impacts of the comparative static results presented above. The welfare impacts of the price changes are reported in Tables 7.10, 7.11, 7.12, 7.13, 7.14 and 7.15 for the cases of coffee, cassava, livestock, maize, millet and sorghum. We show the impacts of shocks to the market structure. To illustrate the complementarities, we show results for a combination of shocks to market structure and international prices

Table 7.10 Coffee price changes and household welfare

	Baseline	Leader split	Leaders merge	Exit of largest	Equal market shares	Perfect competition
Total						
Competition policy	0.00	0.01	−0.01	−0.01	0.05	0.11
International price	0.55	0.55	0.54	0.54	0.60	0.68
Poor						
Competition policy	0.00	0.01	−0.01	−0.01	0.06	0.12
International price	0.58	0.59	0.57	0.57	0.64	0.72
Non poor						
Competition policy	0.00	0.01	−0.01	−0.01	0.05	0.11
International price	0.53	0.54	0.53	0.52	0.59	0.66
Producers						
Competition policy	0.00	0.02	−0.02	−0.03	0.20	0.39
International price	1.95	1.98	1.93	1.92	2.15	2.42

Note: First order impact on household welfare

Table 7.11 Cassava price changes and household welfare

	Baseline	Leader split	Leaders merge	Exit of largest	Equal market shares	Perfect competition
Total						
Competition policy	0.00	0.00	0.00	0.00	0.00	−0.01
International price	−0.14	−0.14	−0.14	−0.14	−0.14	−0.17
Poor						
Competition policy	0.00	0.00	0.00	0.00	0.00	−0.02
International price	−0.31	−0.31	−0.30	−0.30	−0.31	−0.36
Non poor						
Competition policy	0.00	0.00	0.00	0.00	0.00	0.00
International price	−0.03	−0.03	−0.03	−0.03	−0.03	−0.03
Producers						
Competition policy	0.00	0.00	0.00	0.00	0.00	0.09
International price	1.30	1.30	1.28	1.28	1.30	1.50

Note: First order impact on household welfare

Table 7.12 Livestock price changes and household welfare

	Baseline	Leader split	Leaders merge	Exit of largest	Equal market shares	Perfect competition
Total						
Competition policy	0.00	−0.07	0.01	0.10	−0.14	−0.42
International price	−0.13	−0.21	−0.10	−0.01	−0.26	−0.65
Poor						
Competition policy	0.00	−0.01	0.00	0.02	−0.03	−0.09
International price	−0.03	−0.04	−0.02	0.00	−0.05	−0.14
Non poor						
Competition policy	0.00	−0.10	0.01	0.15	−0.22	−0.63
International price	−0.19	−0.31	−0.15	−0.02	−0.40	−0.98
Producers						
Competition policy	0.00	0.36	−0.04	−0.52	0.77	2.26
International price	0.68	1.10	0.53	0.06	1.41	3.49

Note: First order impact on household welfare

Table 7.13 Maize price changes and household welfare

	Baseline	Leader split	Leaders merge	Exit of largest	Equal market shares	Perfect competition
Total						
Competition policy	0.00	0.02	−0.01	−0.03	0.05	0.16
International price	0.12	0.15	0.10	0.09	0.15	0.42
Poor						
Competition policy	0.00	0.03	−0.02	−0.04	0.07	0.23
International price	0.18	0.22	0.14	0.13	0.23	0.61
Non poor						
Competition policy	0.00	0.04	−0.02	−0.05	0.08	0.28
International price	0.22	0.27	0.18	0.16	0.28	0.74
Producers						
Competition policy	0.00	0.20	−0.12	−0.28	0.45	1.59
International price	1.22	1.49	0.98	0.87	1.54	4.15

Note: First order impact on household welfare

(we comment on the results for other complementarities at the end). We also report average results for the total population, the poor, and the non-poor, and producers.

As it is always the case in our analysis, competition leads to positive welfare results (except for producers of import crops), and increases in international prices lead to positive welfare results if the crop is exportable and negative if it is importable. We must note, however that this time two exceptions arise: Cassava and Livestock. We must note that Cassava is not an exportable food crop every

Table 7.14 Millet price changes and household welfare

	Baseline	Leader split	Leaders merge	Exit of largest	Equal market shares	Perfect competition
Total						
Competition policy	0.00	0.00	0.00	0.00	0.00	0.00
International price	0.01	0.01	0.01	0.01	0.02	0.02
Poor						
Competition policy	0.00	0.00	0.00	0.00	0.00	0.01
International price	0.02	0.03	0.02	0.02	0.03	0.04
Non poor						
Competition policy	0.00	0.00	0.00	0.00	0.00	0.00
International price	0.00	0.00	0.00	0.00	0.00	0.01
Producers						
Competition policy	0.00	0.01	0.00	−0.01	0.16	0.28
International price	0.82	0.84	0.82	0.81	0.99	1.17

Note: First order impact on household welfare

Table 7.15 Sorghum price changes and household welfare

	Baseline	Leader split	Leaders merge	Exit of largest	Equal market shares	Perfect competition
Total						
Competition policy	0.00	0.00	0.00	0.00	0.00	0.01
International price	−0.01	−0.01	−0.01	−0.01	−0.01	−0.01
Poor						
Competition policy	0.00	0.00	0.00	0.00	0.01	0.02
International price	−0.04	−0.04	−0.04	−0.04	−0.03	−0.03
Non poor						
Competition policy	0.00	0.00	0.00	0.00	0.00	−0.01
International price	0.01	0.01	0.01	0.01	0.01	0.01
Producers						
Competition policy	0.00	−0.01	0.01	0.01	−0.09	−0.18
International price	0.32	0.32	0.33	0.33	0.23	0.20

Note: First order impact on household welfare

year, but on average. Therefore the survey was probably taken in a year in which Uganda was a net importer of cassava. For Livestock, however, this is probably to an omission in the sample of large producers.

Having given a broad description of the results, we start by describing our only cash crop for Uganda, coffee, in Table 7.10. We note that there are overall welfare gains from competition policy and an increase of international prices, and that the poor benefit slightly more than the non-poor. Moving on to the exportable food crops, we have cassava and livestock in Tables 7.11 and 7.12, and maize and millet, in Tables 7.13 and 7.14. For maize, the poor benefit less from competition policy than the non-poor, while the opposite is true for millet. Finally, we have our only importable food crop, sorghum, whose results we show in Table 7.15. We see that the poor benefit more than the non-poor, and that the non-poor are actually hurt by the decrease in prices induced by an increase in competition. Producers are hurt as well, as expected.

References

African Center for Economic Transformation (ACET) (2015) Uganda country report. Sustainable rural development project, Accra, Ghana

Kraybill D, Kidoido M (2009) Analysis of relative profitability of key Ugandan agricultural enterprises by agricultural production zone. Uganda Strategy Support Program (USSP) Background Paper no. USSP 04

MAFAP (2013) Review of food and agricultural policies in Uganda. MAFAP Country report series, FAO, Rome

Nhassico D, Muquingue H, Cliff J, Cumbana A, Bradbury H (2008) Review: rising African cassava production, diseases due to high cyanide intake and control measures. J Sci Food Agric 88:2043–2049

USAID (2011) Uganda coffee supply value chain analysis: profiling the actors, their interactions, costs, constraints and opportunities. Prepared by Chemonics International Inc.

Vargas Hill R (2010) Liberalisation and producer price risk: examining subjective expectations in the Ugandan coffee market. J Afr Econ 19(4):433–458

Chapter 8
The Case of Madagascar

Our last case study for the first part of the book is Madagascar where we cover rice and vanilla. Next section describes the household survey data. The institutional arrangements for rice and vanilla are briefly presented in Sect. 8.2. Sections 8.3 and 8.4 present the farmgate price simulation results from the model and the household welfare simulations.

8.1 The Household Survey Data

The household data for Madagascar comes from *L'Enquête Periodique auprès des Ménages de 2005* (The Periodic Household Survey of 2005), with more than 54,000 observations corresponding to more than 11,000 households. The sample is evenly split between rural and urban households (51.44 % rural), however the population is not (78 % is rural). As it is usual with the countries under analysis, the population is young, 46 % of the population is 14 or under, and only 2.7 % are 65 or above. The population is evenly split between females and males, females outnumbering males slightly, (females correspond to 50.6 % of the population) but males greatly outnumber females as household heads (females only represent 18.9 % of households heads). Households are large (5.97 people per household on average) and slightly larger in rural areas (6.07 in rural versus 5.62 in urban).

We show the distribution of income in Fig. 8.1, measuring income as per capita household consumption. We show both the rural and the urban distribution of income. As usual, urban households are wealthier, on average, that is the distribution of urban income lies to the right of the distribution of rural income, and more unequal, meaning the urban distribution of income is flatter. Since almost 80 % of the population is rural, the national distribution of income resembles the rural distribution of income more closely than the urban one.

N. Depetris Chauvin et al., *Agricultural Supply Chains, Growth and Poverty in Sub-Saharan Africa*, Advances in African Economic, Social and Political Development, DOI 10.1007/978-3-662-53858-6_8

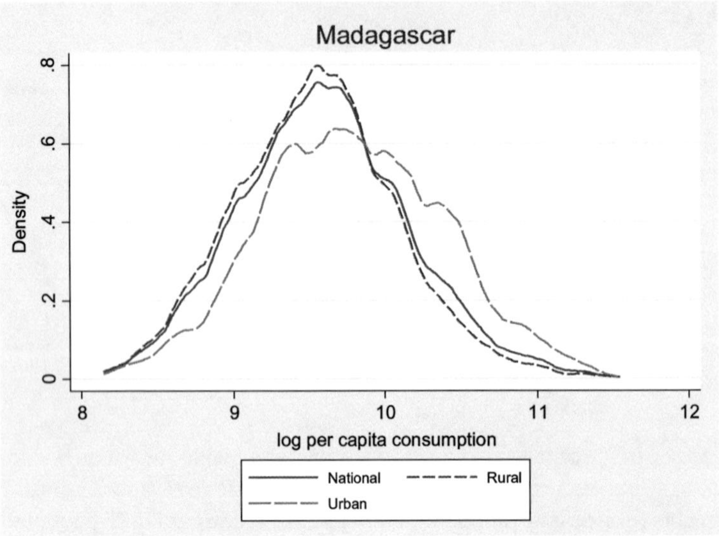

Fig. 8.1 The distribution of income density of (log) per capita household expenditure. Source: Madagascar Household Survey (2005)

We turn to the sources of income and the patterns of consumption across households in Table 8.1. In this table, we report consumption patterns both for urban and rural regions, both cash expenditures and the value of auto-consumption. The latter is more than twice as high as the former in rural areas, which is to be expected. The share of food consumption is also higher in rural areas, but by a smaller margin: 86.8 versus 76.7 %, following Engel's law. Rice is the most important crop by far, amounting to 44 % of total consumption on average. In Table 8.2, we focus on the income side. Again, auto-consumption in rural households more than doubles auto-consumption in urban ones (40.8 versus 17.5 %). Rice reemerges as the most relevant crop, amounting to 30 % of total income.

We now describe the patterns of income and expenditure sources along the income distribution (Figs. 8.2 and 8.3). In these graphs we show the average share of food expenditure and the average share of food income, both including cash expenditures as well as auto-consumption. In food expenditures we see that it remains stable for both rural and urban households, and then falls. Urban households devote a smaller share of their expenditures than rural ones, along the entire income distribution. The share of food income behaves differently; for rural households, it remains high and stable along the income distribution, whereas for urban ones it is declining in income. It is also significantly smaller for urban households: 20 % at the top of the income distribution versus 80 % for rural households. It is evident that food crop production is pervasive for rural households, even for the wealthiest ones. With this data we can already draw a preliminary conclusion: food price increases will hurt the wealthier urban household the most.

Table 8.1 Budget shares

Madagascar	Total	Rural	Urban
Total consumption per capita	100.0	100.0	100.0
Expenditures	58.7	52.9	79.1
Food	43.2	39.7	55.8
Manufactures	11.6	10.2	16.6
Services	3.1	2.4	5.8
Others	0.7	0.6	0.9
Auto-consumption	41.3	47.1	20.9
Auto-consumption food	41.3	47.1	20.9
Auto-consumption others	0.0	0.0	0.0
Total food consumption	84.6	86.8	76.7
Total crops	60.4	63.0	50.9
Maize	2.2	2.4	1.4
Rice	44.5	46.0	39.3
Poultry	0.8	0.8	0.8
Livestock	3.5	3.3	4.2
Cassava	6.7	7.6	3.6
Cowpea	0.2	0.2	0.2
Yam	2.4	2.8	1.3
Cotton	0.0	0.0	0.0

Source: Madagascar Household Survey (2005)

Table 8.2 Income shares

Madagascar	Total	Rural	Urban
Total income per capita	100.0	100.0	100.0
Incomes	64.2	59.2	82.5
Food (agriculture)	33.5	38.0	17.1
Wage	14.2	8.1	36.1
Enterprises	12.9	10.2	22.5
Transfers	3.7	2.8	6.8
Auto-consumption	35.8	40.8	17.5
Auto-consumption food	35.8	40.3	17.5
Auto-consumption others	0.0	0.0	0.0
Total food income and AC	69.2	78.9	34.6
Total crops	56.3	64.1	28.2
Maize	2.5	2.8	1.2
Rice	30.6	35.1	14.5
Poultry	2.2	2.6	1.1
Livestock	10.3	11.4	6.4
Cassava	7.1	8.2	3.4
Cowpea	0.2	0.2	0.0
Yam	2.7	3.0	1.2
Cotton	0.7	0.9	0.3

Source: Madagascar Household Survey (2005)

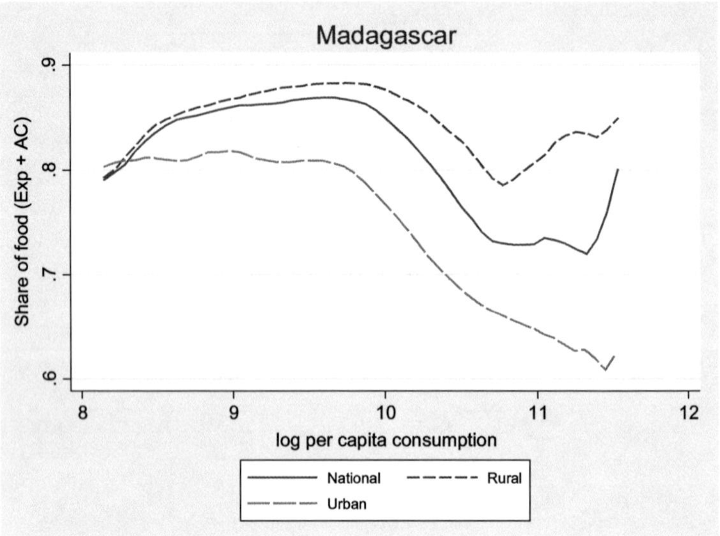

Fig. 8.2 Total food budget share across the income distribution. Source: Madagascar Household Survey (2005)

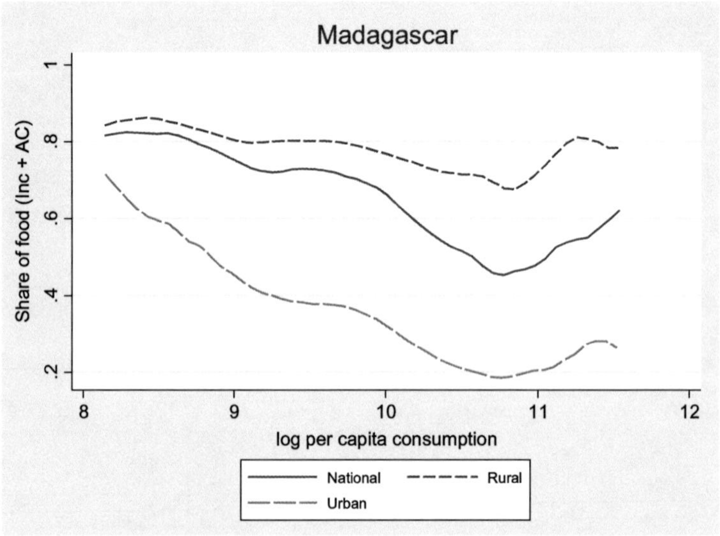

Fig. 8.3 Total food income share across the income distribution. Source: Madagascar Household Survey (2005)

To be able to draw more precise conclusions, we move on to study specific crops (Figs. 8.4, 8.5 and 8.6). We take Rice and Vanilla as our crops under analysis. We start with rice in Figs. 8.4 and 8.5. Rice expenditure and income follow two very

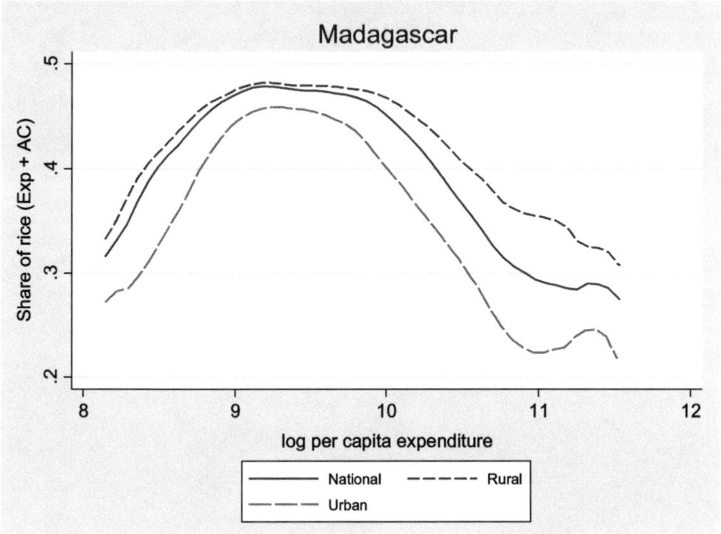

Fig. 8.4 Rice budget share across the income distribution. Source: Madagascar Household Survey (2005)

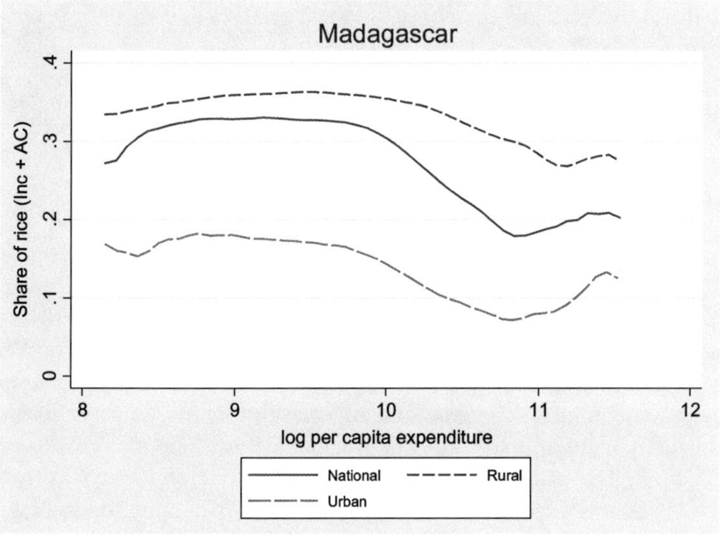

Fig. 8.5 Rice income share across the income distribution. Source: Madagascar Household Survey (2005)

distinct patterns. The former has an inverted u shape, with 50 % of total expenditure at the top, and only 30 % for the poorest and 20 % for the wealthiest households. Urban households and rural households follow the exact same pattern, although rural households have higher shares, as usual. Rice income shares are stable for rural

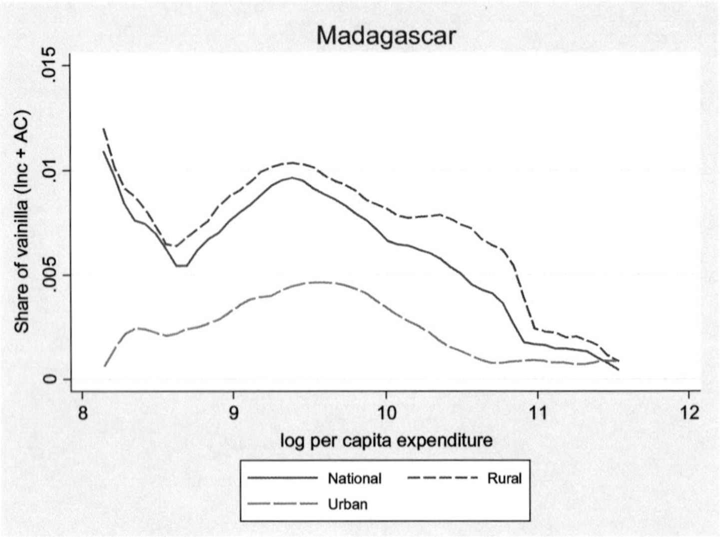

Fig. 8.6 Vanilla income share across the income distribution. Source: Madagascar Household Survey (2005)

households at 35 % to fall slightly for wealthier households to 30 %, whereas urban households follow the same pattern with shares of 20 % and 15 % respectively.

We also show non-parametric regressions of our only cash-crop in Madagascar, vanilla (Fig. 8.6). Vanilla has much smaller shares than rice, making the parametric regression rather unstable. However, we can clearly see that its shares are declining in income for rural households, and that few urban households produce the crop.

8.2 Food and Cash Crop in Madagascar

Madagascar has a unimodal tropical climate characterized by alternating rainy and dry seasons, the lengths of which vary from one region to another. Because altitude accentuates temperature differences, the dry season can thus be particularly cool in the highlands where, sporadically, there can be frost (regions of Antsirabe and Ambatolampy). The East Coast is particularly well watered (over 2000 mm annually over 11 months). On the other hand, the South is notable for its low rainfall (275 mm at Toliary) and a very long dry season. Because of this heterogeneity in ecology, Malagasy farmers are able to grow temperate crops such as apples, pears, plums, grapes and citrus fruits and tropical products such as mangoes and lychees, as well as a wide variety of other crops including coffee, cloves, sisal, maize, rice, and tubers. Agriculture accounts for almost 30 % of GDP, 40 % of export earnings and employs more than 70 % of the labor force. About 5 % of the land area is cultivated at any given time, of which 16 % is irrigated. The average farm size is 1.3 ha, with most farmers practicing subsistence agriculture.

Two important policy that has been guiding the agricultural sector are the Madagascar (2010) and Madagascar (2008). These policies seek to increase agricultural productivity and improve food security in the country. They were designed to contribute to the achievement of several Millennium Development Goals (MDGs), including the eradication of extreme poverty and hunger, gender equality and empowerment of women, as well as environmental sustainability. However, the political instability that the country experienced in 2009 and is still recovering from have stalled the full implementation of these policies. This is why the sector remains a priority for the Government of Madagascar, which allocated 21 % of the public investment budget for 2013 to the sector (AfDB 2013).

We focus on two crops in Madagascar, rice and vanilla (Figs. 8.7 and 8.8). Rice is economically and politically very important in Madagascar. It is the leading crop

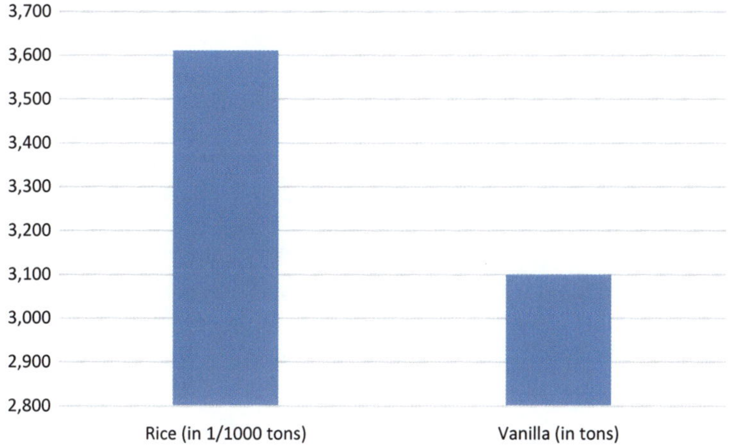

Fig. 8.7 Crops production in Madagascar (in tons, 2013). Source: FAOSTAT Online

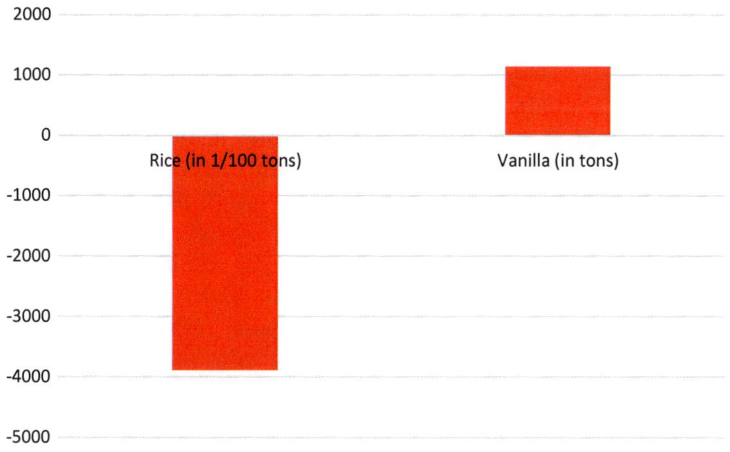

Fig. 8.8 Crops net export in Madagascar (in tons, 2013). Source: FAOSTAT Online

grown by the vast majority of rural households and the main food staple, accounting for 48 % of total calorie consumption. The area under rice increased to about 1.81 million ha in 2010 from 1.15 million ha in 1995. Production increased from 2.45 million tons in 1995 to about 4.74 million tons of paddy rice in 2010. Rice yield increased slightly from 2.1 t/ha in 1995 to 2.6 t/ha in 2010. Enhanced production gradually reduced the need for rice imports. The rice self-sufficiency ratio was about 96.7 % in 2009. There are four principal types of rice growing: irrigated rice, rainfed lowland rice, upland rainfed rice (called tanety), and rice as a first crop after slash and burn (called tavy). In terms of cultivated area, irrigated rice is the most important, covering 82 % of all area under rice in 2008. About 60 % of irrigated rice is transplanted. Rice is grown in six zones of Madagascar: the north, northwest, and central-western regions; the central part of the Malagasy highlands; the east; and the central-eastern part, including Lake Alaotra, with its swampy areas, plains, and valleys suited for rice. Most rice produced in Madagascar is retained for home consumption. Out of total paddy production of around four million tons, only one-quarter to one-third is marketed.

Minten et al. (2006) explored the constraints on agricultural productivity and priorities in boosting rice productivity in Madagascar using a range of different data sets and analytical methods and found that access to agricultural equipment, access to cattle for traction and transport and access to labor are ranked among the top four constraints faced my rice farmers. The second most important set of constraints also identified in Minten et al. (2006) relates to shocks associated with plant disease, drought and flooding. More than half of all households report these constraints to be 'quite' or 'very important'. By contrast, less than 40 % of households identify land tenure insecurity or the siltation of land as important constraints and these are more commonly identified as not a constraint on agricultural productivity.

Producing 60 % of the global supply, Madagascar is the largest world producer of natural vanilla. Its production grew from 525 t in 2003 to about 3100 t in 2013. It is mainly produced by smallholder farmers. Vanilla production thrives in specific locations and conditions (25° north or south of the Equator). It is mostly cultivated in the Sava regions, which consists of Sambava, Antalaha, Vohemar, and Andapa. In 2000, the vanilla production line in Madagascar employed about 20,000 growers and 5000 producers. Madagascar largely exports vanilla to France, Germany, Canada, and the USA. Cultivation of vanilla is difficult and time consuming process that requires a large quantity of labor for planting, harvesting, and curing the produce. This is the reason why vanilla production is less suitable for large scale plantations. Harvested vanilla pass through a complex network of value chain actors in both its uncured and cured form. Vanilla exports used to be monopolized by the government and the associations of vanilla exporters. But today it is dominated by a set of more than 30 exporters. Likewise, the government used to set the final prices, both for the green bean and the final product. But, today prices are determined by international market forces. Vanilla export represents a large portion of government revenues, thus monitoring vanilla production and vanilla trade in the world became some of the priorities of the governments of Madagascar (Alwahti 2003).

Some of the constraint facing the vanilla value chain include, fluctuating prices and margins, lack of resources, poor quality product, and theft. Since the government left the price setting process to market influences, the pricing system of vanilla became less transparent. As a results, margins are distributed unevenly among value chain players to the detriment of farmers. Farmers also face both credit and insurance market failures. They lack resources to expand their operations and given the relatively small number of exporters, farmers have low bargaining powers. Theft is also a major constraint facing the value chain. To combat it, the movement of pods is strictly regulated in Madagascar. For example pods cannot be transported at night, and there are harsh penalties for stealing vanilla. Some farmers also 'tattoo' their beans, stamping a unique mark into them while they are still on the vine.

In Madagascar, there is no marketing board responsible for buying rice neither is there any government involvement in setting rice prices. Rice in most communes of Madagascar are sold privately (individually) on the spot through negotiation between farmers and traders. However, Madagascar is a net rice importing country. We have noted five major importers. Nivoniaina is the top exporter, controlling 17 % of the rice imported in the country. Felana and Scim come in second and third position with a market share of 12 % and 11 %, respectively. In fourth position Cociama and Olam Madagascar control each 9 % of the market share. Price of vanilla is determined by market forces. The elimination of the marketing boards in 1995 substantially reduced government involvement. Cadot et al. (2009) showed how the removal of marketing board and other institutional changes reduced poverty among vanilla farmers. Three exporters control 60 % of the market share. These include Trimera Agro Food with 30 % of the market share, Ramanadraibe and L'Etablissement Germaine with 20 % and 10 %, respectively (Table 8.3).

Table 8.3 Market shares in Madagascar

Vanilla		Rice	
Company	Shares (%)	Company	Shares (%)
Trimera Agro Food	30.00	Nivoniaina	17.00
I'Etablissement Germain	10.00	Felana	12.00
Ramanandraibe Export Co	20.00	Scim	11.00
Remaining 30 exporters	40.00	Cociama	9.00
		Olam Madagascar	9.00
		Others	41.00

Source: Stakeholders interviews

8.3 Simulation Results

In this section, we use our model to perform several simulations consisting of comparative static result stemming from the model. We start with rice. Rice is an importable food crop in Madagascar and therefore we model it as such. The market concentration in rice is relatively low, so we expect small price changes for most simulations. These price changes are presented in Table 8.4. We see in row 1 that convergence to Perfect Competition leads to a decline of prices of 10.51 %, a large response. In row 2, we see a relative small elasticity: a 10 % increase in international prices leads to a 7.67 % rise in local prices. The rest of the shocks in rows 3–9 are small, as usual, none of them lead to a variation in prices above 1 %. An example of complementarities in the results would be the joint effect of Perfect Competition and an increase in the Cash Crop Price, the joint effect being −10.54 and the sum of the effects −9.52 %. An example of substitutability would be the combination of Perfect Competition and rise in International Prices, the joint effect being −1.85 and the sum −2.83 %.

We turn now to the case of Vanilla. Vanilla is a cash crop, and as such, exportable. Its market concentration is higher than in rice, as it is common with cash crops. The largest firm has 30 % of the market, followed by a firm with 20 %. Therefore, we should expect larger price effects, and it is indeed the case. Moreover, since the crop is exportable, the signs of the variations are the opposite. This is

Table 8.4 Farmgate price simulation results for rice

	Baseline	Leader split	Leaders merge	Exit of largest	Equal market shares	Perfect competition
Competition policy	0.00	−0.64	0.59	0.82	−5.16	−10.51
Increase of 10 % in						
International price	7.67	7.07	8.15	8.38	2.57	−1.85
Marginal cost of producing cash crop	−0.17	−0.83	0.37	0.59	−5.31	−10.54
Fixed cost of producing cash crop	−0.28	−0.95	0.24	0.46	−5.44	−10.54
Household resources (endowment)	0.52	−0.19	1.12	1.35	−4.65	−10.54
Risk and food security parameter	0.02	−0.68	0.59	0.83	−5.31	−10.54
Cash crop price	0.99	0.24	1.62	1.85	−4.28	−10.54
Marginal cost of producing food crop	0.25	−0.46	0.83	1.06	−5.02	−10.54
Non-farmer demand	0.19	−0.50	0.75	0.98	−4.99	−10.54

Source: Simulation results from the model of Chap. 2

Table 8.5 Farmgate price simulation results for vanilla

	Baseline	Leader split	Leaders merge	Exit of largest	Equal market shares	Perfect competition
Competition policy	0.00	0.93	−0.68	−1.03	7.55	11.98
Increase of 10 % in						
International price	23.08	24.18	22.14	21.76	31.04	36.98
Marginal cost of producing cash crop	0.72	1.51	0.10	−0.22	8.09	11.98
Fixed cost of producing cash crop	0.22	1.11	−0.44	−0.80	7.75	11.98
Household resources (endowment)	−0.60	0.32	−1.36	−1.71	6.84	11.98
Risk and food security parameter	0.87	1.71	0.23	−0.12	8.40	11.98
Food crop price	0.56	1.39	−0.04	−0.35	7.88	11.98
Marginal cost of producing food crop	−0.15	0.77	−0.87	−1.23	7.45	11.98
Non-farmer demand	0.00	0.93	−0.68	−1.03	7.55	11.93

Source: Simulation results from the model of Chap. 2

what we show in Table 8.5. A convergence towards Perfect competition increases prices 11.98 %, as we show in Row 1 along all the other changes in market structure. In Row 2 we show the effects of a change in international prices. The elasticities are high, 10 % increase in international prices leads to a 23.08 % rise in local prices. As usual, the effects of the change in local parameters, shown in Rows 3–9 are more limited. We can see complementarities and substitutabilities in the results. An example of the former would be the combination of Perfect Competition and a rise of 10 % in International Prices (joint effect 36.98 %, sum 35.08) whereas an example of the latter would be again Perfect Competition and an increase in the Food and Risk Parameter. The joint effect is solely the effect of Perfect Competition since under a model of a perfectly tradeable good all internal shocks impact only quantities and not prices. The combined effect remains then 11.98 %, whereas the sum is 12.85 %.

8.4 Welfare Simulations

We complete our analysis by discussing the welfare and poverty impacts of the comparative static results presented above. These welfare impacts of price changes are shown in Tables 8.6 and 8.7. As usual, we show the impacts of shocks to the market structure, international prices and its combinations and disaggregate the data in poor/non poor and producers.

Table 8.6 Rice price changes and household welfare

	Baseline	Leader split	Leaders merge	Exit of largest	Equal market shares	Perfect competition
Total						
Competition policy	0.00	0.04	−0.04	−0.05	0.32	0.66
International price	−0.48	−0.44	−0.51	−0.52	−0.16	0.12
Poor						
Competition policy	0.00	0.04	−0.03	−0.05	0.29	0.60
International price	−0.44	−0.40	−0.47	−0.48	−0.15	0.11
Non poor						
Competition policy	0.00	0.04	−0.04	−0.05	0.34	0.70
International price	−0.51	−0.47	−0.54	−0.56	−0.17	0.12
Producers						
Competition policy	0.00	−0.15	0.14	0.19	−1.19	−2.43
International price	1.78	1.64	1.89	1.94	0.59	−0.43

Note: First order impact on household welfare

Table 8.7 Vanilla price changes and household welfare

	Baseline	Leader split	Leaders merge	Exit of largest	Equal market shares	Perfect competition
Total						
Competition policy	0.00	0.01	−0.01	−0.01	0.09	0.14
International price	0.26	0.28	0.25	0.25	0.36	0.42
Poor						
Competition policy	0.00	0.01	−0.01	−0.01	0.10	0.17
International price	0.32	0.34	0.31	0.30	0.43	0.51
Nor poor						
Competition policy	0.00	0.01	−0.01	−0.01	0.08	0.12
International price	0.23	0.25	0.22	0.22	0.31	0.37
Producers						
Competition policy	0.00	0.41	−0.30	−0.45	3.32	5.26
International price	10.14	10.62	9.72	9.56	13.63	16.24

Note: First order impact on household welfare

Table 8.6 show the welfare impacts of the price changes in rice. Despite rice being an important crop in consumption and production, the welfare impacts are small because buying positions net themselves with selling positions. We see that competition policies increase welfare for all groups except the producers themselves, since they are in fact beneficiaries of the elevated prices the supplier's oligopoly sets. A rise in international prices hurts all since households are net buyers of the crop, except for producers.

Table 8.7 shows the welfare impacts of the price changes in vanilla. Here an opposite situation takes place. Vanilla, as we have seen in the first section, is on average irrelevant, its shares are very small. However, price changes calculated in the previous section were large, and for those household that do produce the crop, their shares in income are very large: Vanilla producers derive 44 % of their income from vanilla production on average, and spend none of their income on it, therefore that is a net 44 %. That is why convergence to perfect competition leads only to a 0.14 % rise in overall welfare, but a 5.26 % rise in the welfare of the producers, i.e., 40 times higher. This raises the normative question of who the target of the competition policy should be.

References

African Development Bank (2013) South-West region agricultural infrastructure rehabilitation project (PRIASO). Project Appraisal Report

Alwahti A (2003) TED case studies Number 686. American University, Washington, DC

Cadot O, Dutoit L, de Melo J (2009) The elimination of Madagascar's vanilla marketing board, ten years on. J Afr Econ 18(3):388–430

Madagascar (2008) National seed strategy

Madagascar (2010) National rice development strategy

Minten B, Randrianarisoa J, Barrett C (2006) Productivity in Malagasy rice systems: wealth differentiated constraints and priorities. Cornell University Working Paper

Part II
The Role of Non Market Institutions in Sub Saharan Agriculture

Chapter 9
The Role of Grassroots Institutions in Agriculture in Sub Saharan Africa

9.1 Introduction

Grassroots institutions (GRIs) are mechanisms available to smallholder farmers to coordinate activities both horizontally (among members) and vertically (between members and other value chain actors, inside or outside the community). These organizations can reduce transaction costs by creating economies of scale for input supply, technological transfer, or joint marketing, or by facilitating concerted action between farmers (Staal et al. 1998). Established in sub-Saharan Africa (SSA) during the colonial period in the form of cooperatives, GRIs became a means of promoting production and facilitating the collection of cash crops such as coffee, cocoa, tobacco, cotton, and vanilla for export. Although farmers were already organized in groups to address social and community constraints, the colonial period leveraged these indigenous institutions to introduce market orientation functions and formalized them by establishing bureaucratic links with local and central authorities.

The post-independence experience of GRIs was mixed. In countries such as Ghana, governments used GRIs as preferential channels for the provision of credit, often linked to the distribution of agricultural inputs (Hussi et al. 1993; Debrah and Nederlof 2002). Farmers were often coerced, or at best induced, to join GRIs, and membership in GRIs was usually a prerequisite for obtaining credit from the government (Holmen 1990). However, realizing that farmers saw GRIs mainly as a means of obtaining public support rather than a way to promote competitive agribusiness, the government stopped supporting these organizations during the structural adjustment period (Salifu et al. 2010). In other countries, including East African ones, post-independence governments simply neglected these cooperatives, and/or made inadequate use of them, leaving them highly inefficient and weak (Wanyama et al. 2009; Chirwa et al. 2010). Farmers lost faith in their leaders, who were thought to be corrupt, and this diminished their willingness to contribute to

N. Depetris Chauvin et al., *Agricultural Supply Chains, Growth and Poverty in Sub-Saharan Africa*, Advances in African Economic, Social and Political Development, DOI 10.1007/978-3-662-53858-6_9

due to high transaction costs caused by structural constraints such as bad roads, inefficient marketing systems, or limited demand for the product by local consumers or trading partners. In a study conducted in 2008, Barrett reviewed the empirical evidence on smallholder market participation with a focus on staple food grains (i.e., cereals) in eastern and southern Africa. He concluded by suggesting that interventions aimed at facilitating smallholder organization, reducing the costs of inter-market commerce, and, perhaps especially, improving poorer households' access to improved technologies and productive assets, are central to stimulating smallholder market participation. In other words, market access can be optimized if farmers are allowed to organize among themselves and can access complementary services to facilitate their market integration.

To Be a Medium for the Delivery of Development Programs Recent years have witnessed a renewed policy interest in Community-Driven Development (CDD) programs (Binswanger and Nguyen 2005). CDD programs are development initiatives that provide control of the development process, resources, and decision-making authority directly to community groups. This trend is predicated on the premise that interventions at the level of a local community can deliver more effective and equitable development—a view clearly distinguished from the top-down approach, which gave a central entity the responsibility for implementing a development project and has been criticized as elitist (Binswanger and Nguyen 2005). In practice, agricultural CDD interventions are often channeled through GRIs, given their community-oriented nature. Whether effective and equitable development can be efficiently delivered by assisting GRIs is an empirical question. However, a fast-growing set of literature contends that the composition of a GRI will influence its effectiveness in delivering on development programs (Arcand and Fafchamps 2012). If it is composed primarily of local elites, interventions channeled through them are likely to reflect the preferences and interests of these elites. Similarly, if it forms along gender or ethnic lines, its mode of operation is likely to reflect the interests of specific gender or ethnic groups.

Knowing the makeup of GRIs is thus of interest to policy makers concerned with delivering agricultural programs following a CDD framework. From the outset, it is important to know the prevalence of assortative matching among the members of a GRI. It is true that farmers who share similar characteristics are more likely to come together to establish a GRI; member homogeneity is expected to facilitate communication and the alignment of incentives, thus reducing coordination costs. However, homogeneous groups may encounter disadvantages in creating extra-group links, which are usually crucial for properly developing positive spillover. Therefore, in addition to building the capacity of GRI members to negotiate prices, donor's programs ought to encourage some degree of inclusiveness among GRIs if they are expected to facilitate shared benefits within a community. A proper balance of inclusiveness must be advised in order to maximize the benefits to members.

To Allow Farmers to Benefit from the Growing Demand for Quality The rising importance of quality and other types of standards, as well as a higher level of market concentration in the downstream segments of the value chain, is expected to potentially exclude small-scale and/or asset-poor farmers. These farmers generally lack the capacity to respond to the requirements for quality, consistency, volume, and transactional processes demanded by the growing modern food industry (Reardon et al. 2009). Current procurement system changes include: (1) a shift from no standards or public standards to the use of private standards for quality and safety, and (2) a shift from spot market relations in traditional wholesale markets to the use of vertical coordination mechanisms, including explicit or implicit contracts (Reardon and Berdegué 2002). Such trends call for greater attention to improving coordination mechanisms among agents of the value chain in policies and interventions aimed at enhancing the performance and the level of market integration of smallholder farmers (Dorward and Kydd 2004). Coordination at the smallholder level is difficult, given these farmers' atomistic nature and the sometimes high cost of coordination. Nevertheless, while individual farmers have no impact on the value chain alone, they constitute an indispensable force together. Bringing farmers together will allow them greater visibility and responsiveness to the new market incentives.

9.2 The Potential Welfare Effects of GRI Membership

The economic literature provides evidence of the impacts of GRI membership on households'/farmers' welfare. We have identified five areas in which GRI membership could potentially affect household welfare: efficiency, insurance and credit, marketing, technology transfer, and social capital (Fig. 9.2).

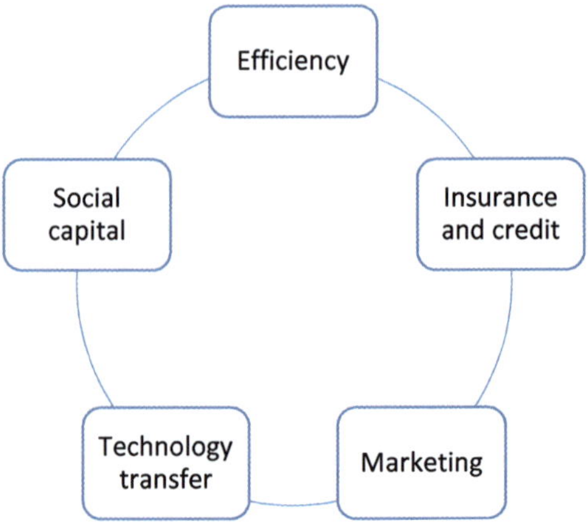

Fig. 9.2 Topics covered in the GRI literature

9.2.1 GRIs and Efficiency

Collective action has often been viewed as a potential channel through which to address bottlenecks in market-oriented production geared towards poverty eradication and the promotion of food security. Makundi et al. (2013) explored the role of collective action in smallholder market engagement among sweet potato producers in southwest Kenya. Makundi et al. found that collective action has the potential to enhance market participation by providing a good platform for the informal exchange of information, which translates into lower transaction costs and thus higher efficiency.

Fischer and Qaim (2011) sought to expand the commonly used concept of farmer group participation by distinguishing between different intensities of participation in a rapidly changing market environment. The results of this study revealed that productivity and technology adoption are positively associated with the intensity of group participation, while distance to collection centers negatively affects group participation—primarily because when the distance is greater, farmers sell their produce to farm gate traders who pay cash on delivery. Furthermore, a larger group size reduces the frequency of participation, possibly suggesting that close social ties enhance group participation.

Ampaire et al. (2013) investigated factors influencing the effectiveness of rural producer organizations in linking members to output markets in Uganda. The results indicated that democratic governance was positively correlated to producer organization effectiveness, and, ultimately, that democratic structures help members by enhancing group participation, primarily in decision-making. A positive correlation between producer group size and increased effectiveness was also demonstrated, due largely to the ability to achieve economies of scale. However, distance to producer groups' meeting site was negatively correlated with the members selling their produce through the organization.

More recently, Macharia et al. (2014) sought to investigate the effect of transaction costs on the intensity of participation among smallholder farmers, and how this translates to an improvement in their ability to participate in formal markets for enhanced incomes and livelihoods. Transaction costs were found to be a major obstacle to smallholders' active participation in formal markets. The study favored policies that encourage the emergence of collective institutions, geared towards the availability of accurate and timely market information.

9.2.2 GRIs, Informal Insurance, and Microfinance

There is a growing call for a reevaluation of the ways in which government and NGOs deliver microcredit, and the new paradigm would aim to leverage rural cooperatives for informal sector financing and insurance. Obademi and Ogbuji (2013) sought to investigate the best approach to correct the deficit in the supply of credit to the informal sector through rural cooperatives, against the backdrop of the underperforming formal financial sector in Nigeria. The study concluded that rural cooperatives are better adapted to addressing key issues in the supply of agricultural credit, including not only the cost of credit but also the timeline of repayment. This is usually facilitated by a cross-guarantee, a procedure often used by participating members.

The World Bank notes that GRIs provide farmers with opportunities for income generation through microfinance schemes, as well as security and protection in times of shock through mutual insurance schemes. Some scholars have also maintained that GRIs allow farmers to better cope with risks, particularly when the private sector, government, civil societies, and donors fail to provide safety nets or insurance against these risks. Drawing from these two arguments, Salifu et al. (2010), in the paper "Farmer Based Organizations in Ghana," postulated that farmers generally organize themselves into FBOs to improve their chances of receiving grants, loans, or training; for collective production and marketing; to share labor; and to engage in other economically beneficial activities.

Further, Hussain (2014) examined the historical development of cooperative organizations in Nigeria, discussing how these rural groups form to protect members against shocks and hazards through mutual insurance. The paper concluded that in Nigeria, rural cooperatives provide locally needed services and employment, while contributing a sense of community and social cohesion. They also allow members to pool their resources to meet individual needs that cannot be resolved by individual members' limited financial capacity.

9.2.3 GRIs and Marketing

Marketing is a major constraint faced by smallholders in rural developing economies. Getting the best price for their produce may require additional costs that farmers cannot afford by themselves. For this reason, leveraging the reduction in transaction costs due to the economies of scale brought in by GRIs will improve farmers' marketing potential. Barham and Chitemi (2009) attempted to identify and understand the underlying factors that enable smallholder producer groups to improve their market situation, and further examined the impact of group characteristics and asset endowments on collective-action initiative to improve the group's marketing performance. The study found that farmers' groups endowed with a core set of natural assets (e.g., a reliable water source or good land and soil) enjoyed a large number of marketing strategy options, while groups lacking these natural assets found their marketing strategies highly constrained. Interventions in the form of direct market linkages aided in group market success.

Shiferaw et al. (2009) analyzed the role of institutional and organized innovations in improving the performance of rural markets in less favored areas. The paper defined "less favored areas" as those with good agricultural potential but limited access to markets. It showed that collective marketing groups get better prices for their members and that the incentives for joining collective marketing groups seem higher for smallholders who face high transaction costs.

Fischer and Qaim (2012) provided insights into the impact of marketing and non-marketing outcomes on farmers' collective action in Kenya. The findings from the article suggest that output price advantages associated with farmers' collective marketing groups are positive, but relatively small, contributing to a growing trend in which members of these groups continue to sell individually.

Mujawamariya et al. (2013) explored double side-selling in cooperatives in Rwanda, in which cooperative members tend to sell to traders and non-members sell to cooperatives. This arrangement arose due to the loyalty, trust, and informal contracts between farmers and traders as a direct consequence of the apparent absence of structured mechanisms from the cooperatives' side. The study proposed that it is easier for farmers to sell to traders because they get paid immediately, and because personal contacts between farmers and traders reduce transaction costs. On the other hand, cooperatives can be more attractive because of the benefits members can accrue through favorable marketing and production.

Finally, Agbo et al. (2013) proposed a theoretical model based on agricultural cooperatives with direct selling. The paper studied the market structure generated by the sale of goods through cooperatives conducted simultaneously with direct sales to consumers in a local market. The theoretical model revealed that direct selling and marketing through an agricultural cooperative might benefit each party; cooperatives may not have market power, but can exert an anti-competitive effect on the direct selling market. On the other hand, direct selling can create healthy competition and stimulate farmers to increase production, which could benefit the cooperative.

9.2.4 GRIs and Technology Transfer

Many academic works have identified GRIs as a key vehicle in championing the cause of research and extension delivery in Africa, while others view the impact of GRIs as largely unproven due to a lack of empirical analysis (Rogasa and Golan 2012). Verhofstadt and Maertens (2013) analyzed the impact of cooperative membership on agricultural performance in rural households in Rwanda and concluded that collective action leads to the adoption of modern inputs, increased commercialization of farm production, and higher revenue and incomes. Further evidence supports the idea that agricultural cooperatives promote the transformation of smallholder farms from subsistence-level operations to intensified commercial units.

Agricultural cooperatives provide inputs and linkages to extension services and service providers; thus, their members are expected to be more technically efficient than non-members. Abate et al. (2013) explored the level of technical efficiency of agricultural cooperative members versus that of non-members and found that cooperatives are effective in providing support services that contribute significantly to members' technical efficiency. Therefore, promoting agricultural cooperatives as a complement to public extension services should enhance technical efficiency. Asibey-Bonsu (2012) analyzed FBOs engaged in the provision of services to producers and others in the Ghanaian agricultural supply chain. This study demonstrated that the development of grassroots organizations contributes significantly to improved access to essential agricultural services; in this context, grassroots farmers' organizations could be a most effective marketing channel for agricultural inputs and products, as well as extension delivery.

Boyana and Tshuma (2013) sought to explore the possibility of alleviating food insecurity in South Africa by empowering rural agricultural cooperatives. The paper looked at building capacity among smallholder farmers united through collective action, using extension services and training in basic critical farm business operations. This, coupled with a public-private approach in extension delivery, would go a long way towards shoring up the efforts of agricultural cooperatives to better deal with food insecurity.

9.2.5 GRIs and Social Capital

The impacts of social capital on welfare outcomes by enhancing agrarian cooperatives cannot be overemphasized. Many recent studies have looked at this phenomenon in various configurations. Ruben and Heras (2012) attempted to measure the differences in social interaction between cooperative members on the basis of social norms, trust and reciprocity, and how members are rooted in internal and external networks. The study postulates that those cooperatives whose members share more social capital are better able to build a communitarian response to critical external constraints, present stronger resistance to adversity, and are more capable of recovering access to resources. The communitarian response, however, depends on the kind of social capital available; intra-community bonding social capital is more practical in collective action.

Faysse et al. (2012) sought to determine the extent to which members of local farmers' organizations are geographically spaced. In this context, farmers' organizations are institutions in which farmers talk to each other about agricultural practices and obtain the information they need to address problems related to their own farming practices. The study found that the success of collective action can be enhanced by local discussion networks, with the caveat that this action must be led by a well-functioning local farmers' organization.

Deng and Hendrikse (2013) explored a theoretical analysis of social capital in cooperatives by presenting a model of the interaction between a cooperative's social capital and its economic incentives, and this interaction's influence on product quality. The findings suggested that when the cooperative's level of social capital is high, this social motivation can guarantee high product quality even while economic incentives are weak. On the other hand, as the level of social capital

collective action. This inefficiency led to the disbandment of GRIs during the structural adjustment period, under the auspices of the Bretton Woods institutions, to give way to the free market. While justified on efficiency grounds, the dissolution of these institutions, regardless of the post-independence experience, led to incompleteness in the market and cut smallholder farmers off from the rest of the economy, as most African countries did not have the right market infrastructure to support a proper response to market incentives.

It was not until the early 2000s that SSA witnessed an emergence of both governmental and nongovernmental initiatives seeking to promote the development of GRIs. Renamed farmer-based organizations (FBOs), the remaining agricultural cooperatives were liberalized, detached from government interference and connected to value chain actors. While the function of FBOs remained more or less the same as that of the cooperatives of colonial times, the new name stressed the fact that, unlike cooperatives, FBOs are member-owned, bottom-up, demand-driven, and service-oriented organizations, with no direct connection to government. The motivations for this reemergence were several. They included:

- To improve farmers' market participation (For both input and output),
- To be a medium of development program delivery.
- To allow farmers benefit from the growing demand for quality.

We expand below on the three main arguments that have been brought out for reinvigorating GRIs for Africa's agricultural sector (Fig. 9.1).

To Improve Farmers' Market Participation Early studies have shown that the availability of markets and the proximity of cities influence productivity in agriculture (Goertz 1992). Combined with the low marketed surplus of smallholder farmers, this evidence has prompted a series of analyses of the determinants of market participation (Goertz 1990; Fafchamps and Hill 2005; Mulangu 2011). A seminal study by de Janvry et al. (1991) shows that transaction costs (influenced by the efficiency of markets) affect the marketed surplus rate. A market may be limited

Fig. 9.1 Case for GRIs in Africa

declines, the cooperative must adopt an income rights structure with stronger quality incentives in order to maintain product quality.

Meanwhile, Ntale (2013), in the article "Forcing the dumb to sing? Rejuvenating farmer cooperatives in Uganda amidst waning social capital," attempts to question the orthodoxy, logic, and anticipated success of a collectivization program, which bulks smallholders' lands into collective farms, in the face of a decline in or absence of social capital in Uganda. He argued that societies with a low degree of social trust are less likely to create the kind of flexible and prosperous business organizations—and, for that matter, collective organizations—that can compete in both national and regional economies.

9.3 GRIs and Collective Action Theory

The literature around collective action parallels that surrounding social capital. Uphoff and Wijayaratna (2000) highlighted how structural forms of social capital (i.e., roles, rules, procedures, and social networks) facilitate mutually beneficial collective action, and how cognitive forms of social capital (i.e., norms, values, attitudes, and trust) are conducive to mutually beneficial collective action. They further showed how these forms of social capital brought about successful collective action measures in the management of irrigation schemes in Sri Lanka. Other studies, such as Pretty and Ward (2001) and Krishna (2002), have similarly shown how human and social capital formation—often represented in GRIs—have been pivotal in solving many community development problems, particularly in the area of natural resource management.

The growing literature on social network analysis has brought to the surface the collective nature of the benefits of social capital, as illustrated in the discussion presented in Chap. 10. This evidence was built on the theoretical deliberations of Olson's seminal 1965 model, which triggered the literature of collective action. The model nests into the prisoner's dilemma model, in which collusion between the two prisoners leads to better outcomes than no collusion. While this model has since evolved into a version that allows for multiple agents choosing from more than two options in a dynamic setting, the basic deduction remains the same.

9.3.1 Olson's Model of Collective Action

We illustrate the collective action problem using Olson's model. Drawing upon the framework proposed by Alesina et al. (2001) and Reuben (2003), the basic principle of Olson's model is that when the decision to provide a collective good is analyzed from the individual point of view, there is a high incentive to free-ride on the efforts of others and thus provide sub-optimal goods. Overcoming the free-rider problem is a key characteristic of successful groups.

Suppose a group consists of $N = \{1, \ldots, n\}$ individuals, all of whom can produce a non-negative amount of a certain collective good. More specifically, every individual $i \in N$ produces an amount, $\gamma_i \geq 0$. Thus, the total amount produced of the collective good, Γ, is given by the sum of individual contributions, $\Gamma = \sum_i \gamma_i$. The utility gained by an individual, i, from the consumption of the collective good is captured by $v_i(\Gamma)$, where $v_i' > 0$ and $v_i'' \leq 0$. Although the utility gained by the individual depends on the total amount produced of the good, each individual can value the consumption of the collective good differently. The utility gained by the whole group, $V(\Gamma)$, is given by the sum of the utilities of the individuals, $V(\Gamma) = \sum_i v_i(\Gamma)$. The cost, or utility loss, of each individual for the production of the good is given by the cost function $C_i(\gamma_i)$, where $C_i' > 0$ and $C_i'' > 0$. Finally, in order to formalize the maximization problem faced by each individual, we should note that an individual produces a positive amount of the collective good only if the utility she gains by doing so exceeds the utility she loses, that is, if $v_i(\Gamma) > C_i(\gamma_i)$. Thus, each individual in the group faces the following maximization problem:

$$\max_{\gamma_i} v_i(\Gamma) - C_i(\gamma_i) = v_i\left(\gamma_i + \sum_{i \neq j} \gamma_i\right) - C(\gamma_i)$$

$$\text{s.t. } v_i(\Gamma) > C_i(\gamma_i)$$

In the simplest case, in which each individual takes the actions of the others as constant (that is, when $\frac{d\gamma_j}{d\gamma_i} = 0$ for all i and $j \neq i$), the best response, γ_i^*, of each individual, $i \in N$, is given implicitly by

$$\frac{dv_i\left(\gamma_i^* + \sum_{i \neq j} \gamma_j\right)}{d\gamma_i} = \frac{dC_i(\gamma_i^*)}{d\gamma_i} \text{ if } vi(\Gamma^*) > Ci(\gamma_i^*) \text{ and } \Gamma^* = \sum_i \gamma_i^*$$

$$\gamma_i^* = 0 \quad \text{if } vi(\Gamma^*) \leq Ci(\gamma_i^*)$$

The solution to the problem is obtained when all individuals use their best responses. That is, for every i,

$$\frac{dv_i(\Gamma^*)}{d\gamma_i} = \frac{dC_i(\gamma_i^*)}{d\gamma_i} \text{ if } vi(\Gamma^*) > Ci(\gamma_i^*) \text{ and } \Gamma^* = \sum_i \gamma_i^*$$

$$\gamma_i^* = 0 \quad \text{if } vi(\Gamma^*) \leq Ci(\gamma_i^*)$$

Although this model is extremely simple, it already gives us important insights into the collective action problem. As expected, the individual will produce the collective good up to the point at which her marginal gain equals her marginal loss.

Unfortunately, the individual does not take into account the effect of her production on the utility of other group members, and thus produces a sub-optimal (from the group's perspective) quantity of the good. To see this more clearly, consider the maximization problem the group would face if it behaved as a unitary actor. In this case, the maximization problem would be

$$\max_{\gamma_i \cdots \gamma_n} V(\Gamma) - \sum_i C_i(\gamma_i)$$
$$s.t. V(\Gamma) > \sum_i C_i(\gamma_i)$$

The optimal amount produced by every individual, i, would be given implicitly by

$$\begin{matrix} \sum_j \frac{\partial v_j(\Gamma^{**})}{d\gamma_i} = \frac{dC_i(\gamma_i^{**})}{d\gamma_i} & \text{if} \quad \text{vi}(\Gamma^{**}) > \sum_i C_i(\gamma_i^{**}) \\ \gamma_i^{**} = 0 & \text{if} \quad \text{vi}(\Gamma^{**}) \leq \sum_i C_i(\gamma_i^{**}) \end{matrix} \quad \text{and } \Gamma^{**} = \sum_i \gamma_i^{**}$$

where γ_i^{**} is the amount of the collective good produced by i that maximizes V(Γ). Since $\frac{dv_i(\Gamma)}{d\gamma_i} < \sum_j \frac{\partial v_j(\Gamma)}{d\gamma_i}, \forall \Gamma > 0, i \in N$, and $C_i'' > 0 \, \forall \, i \in N$, it follows that γ_i^{**} must be strictly bigger than γ_i^{*} for all i, and consequently Γ** is strictly bigger than Γ*. In other words, the collective good is undersupplied.

Ostrom (1994) took the concept of collective action further by incorporating social capital. From the outset, she argued that social capital is the arrangement of human resources to improve flows of future income, and that it is created by individuals spending time and energy working with other individuals to find better ways of making possible the achievement of certain ends that, in its absence, would not be possible (Coleman 1966). Farmers' investment in social capital takes the form of bargaining over which rules will be adopted to allocate the benefits and costs of collective action. One underlying assumption that is often made when analyzing the role of collective action is that farmers, or GRI members, are homogeneous. For that reason, evaluations of the impact of GRIs on farmers tend to take the form of a comparison of GRI members and non-members and use econometric analysis to attribute the difference to GRIs. The reality is that GRI participants are often heterogeneous, even if their differences are not always visible. When substantial heterogeneities of capacity and interest exist, the rules benefit and cost sharing will affect the distribution of outcomes among farmers.

9.3.2 Sources of Limitations Among GRIs

In practice, collective action has not always brought effective welfare improvements, as derived by the theoretical deliberations above. Recent results from Chiodi and Mulangu (see Chap. 10) found that while GRI membership improves access to both credit and inputs, it reduces access to the external land market in Ghana. GRIs also face substantial challenges in both managerial and social domains. An important number of these challenges can be characterized as structural "tensions" between different goals or functions, as Muradian (2013) puts it. The ability to cope with these challenges is a key factor influencing GRIs' performance. Some of these tensions have been clearly presented by Bernard and Spielman (2009) and Muradian (2013), among them:

- Social versus business functions. The most prominent trade-off faced by GRIs lies in balancing social/developmental and marketing functions. GRIs tend to be naturally strong in facilitating social services for their members, but weak on marketing because it requires a lot of energy and capacity. Trade-offs between social or developmental functions and marketing functions are well reported in the literature, as deduced from the conclusions of Bernard and Taffesse (2012) and Bernard et al. (2008).
- Membership homogeneity and external links. Most GRIs tend to have a homogeneous membership. The work of Fafchamps and La Ferrara (2012) found evidence of assortative matching among members of a self-help group in Kenya. This homogeneity is expected to facilitate communication and the alignment of incentives, thus reducing coordination costs. However, homogeneous groups may encounter disadvantages in creating extra-group links, which are usually crucial for proper development of marketing functions (Muradian 2013).
- Coordination costs and economies of scale. The creation of economies of scale is one of the key mechanisms through which farmers' groups can enhance the collective bargaining power of small-scale producers, as the Olson (1965) model illustrates. However, as the size of a GRI increases, economies of scale can be undermined by the growing cost of coordinating the activities of a large number of members. Larger groups require more delegation of management tasks to the board of directors or the managers, increasing the probability of principal–agent problems (Levi and Davis 2008).
- External support and autonomy. After reviewing collective arrangements for integrating smallholder farmers into agricultural markets in eastern and southern Africa, Poole and de Frece (2010) concluded that "most successful cases of collective enterprise creation" have depended on a substantial degree of intervention from NGOs and international donors. This is also the case for GRIs in Ghana, as discussed in Salifu et al. (2010), as most GRIs operating in Ghana are dormant and exist only to respond to donor or government initiatives. External intervention can create an environment of expectations by GRI members such that they no longer see the benefit of putting out effort. The challenge, then, is to

find the right balance between external support and sufficient autonomy in cooperative development.

- Meeting standards and satisfying members. Several studies have shown positive synergies between certification schemes and collective action, as illustrated by Muradian (2013), in facilitating market integration and ensuring better prices for smallholder farmers. Kuapa Kokoo for cocoa in Ghana and Oromia for coffee in Ethiopia, both of which are qualified for fair trade export, are good examples. Nonetheless, meeting strict standards normally entails conflict with those members who are not able to deliver products according to the specifications. Exclusion of some members can reduce trust between members and managers, leading to weakened commitment by members and a diminished sense of group identification, which will negatively affect the cooperative's performance (Hernández-Espallardo et al. 2013; Nilsson et al. 2012).

Addressing these tensions has become an objective of NGOs and/or government as a way to help improve the functioning of GRIs. Interventions lie in three broad areas of action: marketing, knowledge and technology transfer, and social capital and managerial skills improvement. These three areas of actions are reflected in the Agricultural Green Revolution for Africa (AGRA)'s

9.4 Mapping and Discussion of Agricultural GRIs in Africa

GRIs in Africa have seen enormous growth and development in recent years, and their contribution to food security cannot be overemphasized. Agricultural cooperatives in SSA countries have gained recognition from local and international organizations across the globe, leading to a proliferation of umbrella organizations in various countries aimed at regulating their activities. For the purposes of this paper, this section will attempt to discuss the general characteristics of GRIs in SSA, stressing their composition or type and their sub-regional share.

A desktop survey was conducted on the different GRIs operating in Africa, seeking to identify the GRIs, their locations, and some elements regarding their characteristics, using a sample of 552 GRIs from about 20 countries. The survey found that there are now tens of thousands of local GRIs in sub-Saharan Africa. They provide services to their members and defend members' interests with other stakeholders, including the authorities responsible for agricultural policy, their commercial partners and development NGOs. GRIs create many basic unions and local federations, which join to form national umbrella organizations.

Agricultural GRIs in the SSA region are diverse in their composition and activity. They usually consist of production/marketing organizations, which form about 41 % of the total; input procurement and credit organizations, constituting about 16 %; and lobbying organizations, which make up 12 %. The remainder is made up of production organizations (11 %), export organizations (8 %), marketing (4 %), and processing organizations (5 % of FBOs). The last group, for purposes of

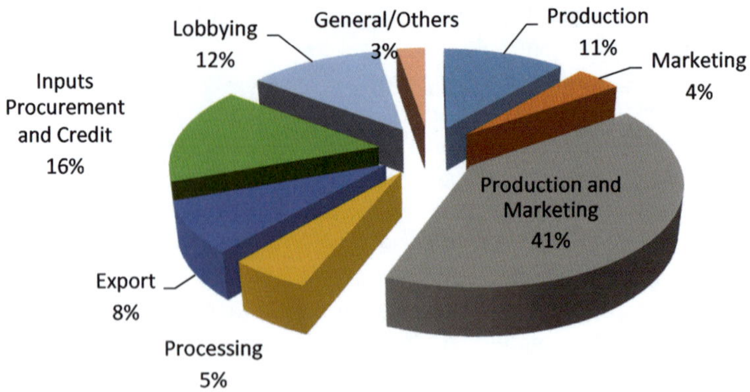

Fig. 9.3 Activities of farmer-based organizations across SSA. Source: Author's construct from GRIs in SSA

this discussion, will be termed "general organizations" and represent about 3 % of FBOs in SSA (see Fig. 9.3).

The production and marketing organizations in SSA cultivate and market their produce collectively and are involved in the provision of a ready market for their members, creating value addition for better profits, capacity building, and input acquisition.

Input procurement and credit organizations facilitate access to credits and inputs, provide legal assistance to their members, and help members acquire inputs and market their agricultural produce. They also provide links to extensive services and agricultural credits and serve as a source of insurance for their members.

Production groups provide their members with a platform to improve and increase their yields through collective farming, and assist farmers in the procurement of inputs and credits, as well as the provision of storage facilities for their produce. Some production organizations are also involved in the manufacture of simple farm equipment.

Marketing organizations help members improve their market outcomes by promoting commodity trading and the provision of a ready market for out-grower schemes, along with identifying markets and creating value addition to improve profits and reduce transaction costs. Further, they improve market accessibility, marketing produce independently to fair trade markets, and help farmers obtain more bargaining power.

The processing- and export-based organizations in SSA are mostly involved in post-harvest activities, notably schemes to facilitate collective processing and exports. These organizations enhance export routes, provide facilities for processing, and are involved in the promotion of export commodities while enhancing the capacities of their members.

GRIs involved in lobbying essentially advocate and represent the interests of farmers, helping farmers access local high markets and negotiating prices for their produce. They also give a strong voice to farmers in all agricultural matters: land acquisition, agricultural funding, market access, and extension.

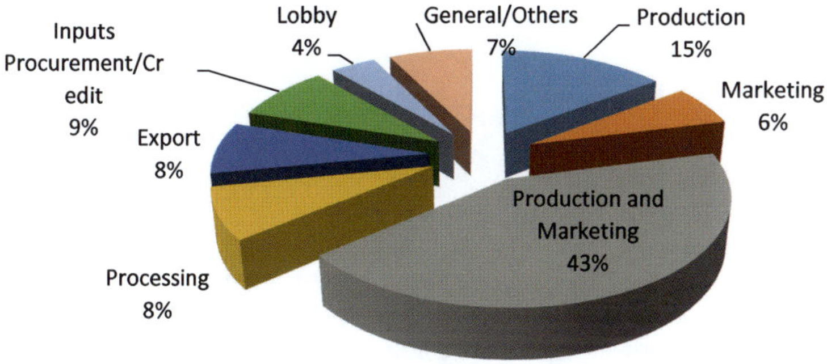

Fig. 9.4 Activities of farmer-based organizations in West Africa. Source: Author's construct from GRIs in SSA

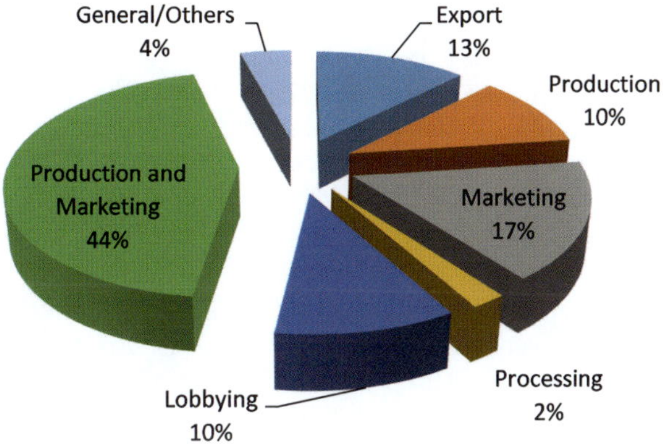

Fig. 9.5 Activities of farmer-based organizations in Southern Africa. Source: Authors' construct from GRIs in SSA

The general organizations are not focused on a single activity, but may perform some or most of the functions above. These organizations are not rigidly defined, but can transform at any point in time to meet the needs of their members.

The dominant types and activities of GRIs also differ from one sub-region to another. In West Africa, the production and marketing organizations dominate, accounting for about 43 % of all GRIs. Production, input procurement and credit, and processing organizations hold shares of 15 %, 9 %, and 8 %, respectively. Export, marketing, and lobby groups make up approximately 8 %, 6 %, and 4 %, respectively, whereas general organizations form 7 % of all GRIs (see Fig. 9.4).

The southern African sub-region has a diverse GRI makeup, tending to mirror that of West Africa. Here, production and marketing groups form about 44 % of GRIs. Marketing, export, and production groups make up 17 %, 13 %, and 10 %, respectively, while lobbying, processing, and general cooperatives hold shares of 10, 2, and 4 %, in that order (see Fig. 9.5). Input/credit groups are not found here, so constitute <0 % of the sample.

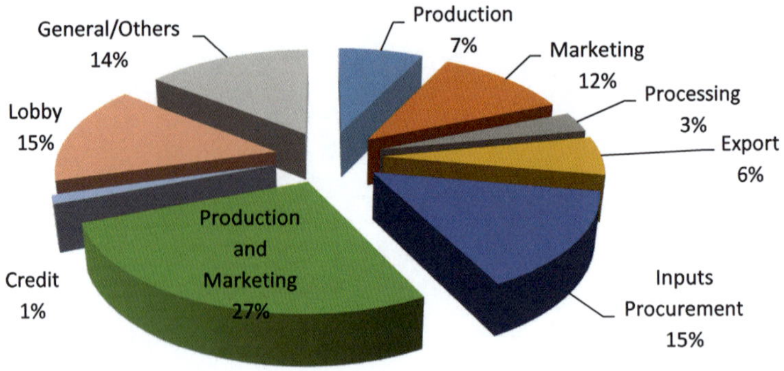

Fig. 9.6 Activities of farmer-based organizations in Eastern Africa. Source: Authors' construct from GRIs in SSA

The East African sub-region is also dominated by production and marketing organizations, which account for about 27 % of GRIs. The lobbying, input procurement, and marketing groups hold shares of 15 %, 15 %, and 12 %, respectively, while general, export, and processing organizations hold 14, 6, and 3 %, in that order. Credit organizations, however, constitute a minuscule 1 % of GRIs in that region (see Fig. 9.6).

In sum, GRIs across sub-Saharan Africa are diverse in their nature and activities; however, production and marketing organizations hold the largest share, followed closely by those involved in input procurement and credit and organizations involved in production only. Through collective action, these organizations are able to achieve economies of scale in production, reduce their transaction costs, and gain market information; this has great implications for food security and sustainable livelihoods.

9.5 Initiatives Led by Donors and Government to Introduce or Revive GRIs

Smallholder farmers in Africa continue to receive support from donor partners. Many of these grassroots farmers' organizations are joined to national umbrella organizations, which have established five regional networks in Africa. These networks include: Platform of Peasant Organizations of Central Africa (PROPAC), Eastern Africa Farmers Federation (EAFF), the Southern African Confederation of Agricultural Unions (SACAU), Network of Peasant Organizations and Producers in West Africa (ROPPA), and the Maghreb Farmers' Union (UMAGRI) in North Africa. The United Nations General Assembly passed a resolution declaring 2012 the "International Year of Cooperatives," in order to showcase the contribution of cooperatives to the socioeconomic well-being of their members. In much the same vein, the African Union declared 2014 to be the "Year of Agriculture and Food

Table 9.1 Support by donor partners to GRIs in Africa

Program title	Duration	Location	Funding agency	Value
Food Facility program	2009–2011	Various	EU	232€ million
ACP Support Program on Agriculture	2007–2011	Various	EU	Unspecified
Improvement of Food Security in Cross-Border Districts	2006–2013	Burundi, Rwanda, Uganda, and Democratic Republic of Congo	The New Partnership for Africa's Development (NEPAD)	Unspecified
National projects on Food Security through Commercialization of Agriculture	2008–2012	Various	UN Food and Agriculture Organization (FAO)	Unspecified
Cassava Initiative for Vulnerable Smallholders	2009–2011	Central and eastern Africa	FAO	Unspecified
Building Commodity Value Chains and Market Linkages for Farmers' Associations	2008–2011	Mozambique	FAO	Unspecified
Livelihood Support Program	2011–2012	Somalia	FAO	Unspecified

Source: Authors' construct using FAO report, 2012

Security." During the past two decades, GRIs in sub-Saharan Africa have seen various interventions from international organizations through their governments or regional networks (Table 9.1).

- Other initiatives by international organizations also seek to improve food security through GRIs. The Alliance for a Green Revolution in Africa (AGRA) has initiated the Market Access Program (MAP), which ensures value addition for farm produce through post-harvest service delivery. AGRA's MAP has aided various cooperatives throughout Africa. In 2010, AGRA provided funding to the Clinton Hunter Development Initiative (CDHI) to improve the post-harvest activities of a soybean producers' organization in Rwanda. The project spanned 2 years, from 2010 to 2012. Farmer Organization Support Centers in Africa (FOSCA), a program established by AGRA in 2011, offers smallholders opportunities to increase income and improve market access and linkages to technical services. Since its inception, over 1880 farmers' organizations have been trained in group governance and management.
- The Support to Farmers' Organizations in Africa Program (SFOAP) started as a pilot program between 2009 and 2012, aimed at building the capacity of smallholder organizations in sub-Saharan Africa and their regional and pan-African networks and support programs, with the overall goal of improving agriculture, rural development, and food security in Africa. The program, supervised and financed by the EU and the International Fund for Agricultural

Development (IFAD), with a total budgetary allocation of approximately 6.2€ million, assisted 55 national farmers' associations in 39 African countries.

- The main phase of the SFOAP, begun in 2013 and expected to end in 2017, intends to build on the successes and lessons learned from the pilot phase to support institutional capacities and build on FBOs' economic services. It also leans towards facilitating the integration of smallholder farmers into the value chain process. A total of 19.9€ million has been allocated for this phase, funded by agencies including the EU, the Swiss Agency for Development and Cooperation (SDC), Agence Française de Développement (AFD), and IFAD. The Food and Agriculture Organization of the United Nations (FAO) and the Technical Centre for Agricultural and Rural Cooperation (CTA) provide additional technical support for the program. The Farmers Fighting Poverty Africa program (FFP/AFRICA), a complement to the SFOAP, is aimed at improving the livelihoods and food security of rural producers in Africa by working with farmers' organizations to help them become more stable and accountable. FFP/AFRICA is managed by AgriCord, a network of nine Organization for Economic Cooperation and Development (OECD) countries: Belgium, Canada, Finland, France, Italy, Netherlands, Portugal, Spain, and Sweden. The 2013–2015 FFP/AFRICA is financed by the EU and AgriCord, with technical support from IFAD, at a total cost of 20.2€ million.
- The Root Capital Program is a social investment fund that aims to spread rural prosperity in poor, vulnerable countries. They work with farmers' cooperatives in over 30 low-income countries, with the intent of helping farmers move from agrarian to commercial food production by gaining access to global supply chains, translating into increased income and food security. This program, financed by the International Finance Corporation (IFC) and the Global Agricultural and Food Security Program (GAFSP) with a budgetary allocation of $5 million, began in 2012 and is scheduled to end in 2016.
- Another initiative to support GRIs in Africa is a grant program named "Strengthening capacity of East African farmer organizations through knowledge management and institutional development" which aimed at strengthening the capacity of EAFF, the umbrella organizations for all FBOs in East Africa. This program, which intends to build capacity through knowledge management and institutional development, commenced in 2012 and is expected to end in 2015; it is sponsored by IFAD, with a budget in excess of $1.5 million. The goal of the program is to mobilize, retain, and deliver services to members of EAFF, with emphasis on three countries: Burundi, Kenya, and Tanzania. It further aims to increase membership in national farmers' organizations and improve farmer-based learning and innovation.
- TechnoServe is another organization involved in the strengthening of FBOs in sub-Saharan Africa. It has worked on a number of projects aimed at creating FBOs and strengthening already-established ones. In 2008, TechnoServe secured a 4-year grant, totaling $47 million, from the Bill and Melinda Gates Foundation (BMGF) to help smallholders benefit from rising demand. This gave rise to the Coffee Initiative, launched by TechnoServe to help raise the incomes

of East African farmers. The goal of this initiative was to help coffee growers in this region organize and establish cooperatives to gain better bargaining power for their produce. The initiative also assisted the cooperatives in establishing processing facilities so they could process their produce collectively.

- The Agriculture Sector Development Program, a ¥34.43 million ($0.28 million) initiative of the Japan International Cooperation Agency (JICA), enables smallholder farmers to get better access to technology systems and infrastructure, translating into improved productivity. This initiative, which started in 2013 and is scheduled to end in 2017, has seen improvement in irrigation for smallholder groups in Tanzania involved in rice production. JICA has also undertaken other initiatives in the rice sector in Africa to boost smallholder organizations, such as the Project for Promotion of Farmers' Participation in Irrigation Management, a JICA initiative begun in Ghana in 2004 and ended in 2006. A total of ¥250 million ($2.07 million) was invested in this program to boost rice cultivation among farmers' groups through irrigation and its management.

The discussion above is not limited to the above interventions, as there are other initiatives by international organizations aimed at improving and equipping GRIs in sub-Saharan Africa. Most of these initiatives seek, knowingly or unknowingly, to address some of the tensions mentioned in Sect. 9.2 by improving marketing and facilitating knowledge and technology dissemination among GRI members.

References

Abate G, Francesconi G, Getnek K (2013) Impact of agricultural cooperatives on smallholders' technical efficiency: evidence from Ethiopia. European Research Institute on Cooperative and Social Enterprises (Euricse) Working Paper no. 50/13

Agbo M, Rousselière D, Salanié J (2013) A theory of agricultural marketing cooperatives with direct selling. Group d'Analyse et de Théorie Economique (GATE) Working Paper 1331

Alesina A, Angeloni I, Etro F (2001) The political economy of international unions. NBER Working Paper 8645, National Bureau of Economic Research

Ampaire E, Machethe C, Birachi F (2013) The role of rural producer organizations in enhancing market participation of smallholder farmers in Uganda; enabling and disabling factors. Afr J Agric Res 8(11):963–970

Arcand JL, Fafchamps M (2012) Matching in community-based organizations. J Dev Econ 98 (2012):203–219

Asibey-Bonsu P (2012) Farmer's organizations in West and Central Africa: high expectations, hard realities. Foundation for Agriculture in Rural Communities (FARM), Ghana country report

Barham J, Chitemi C (2009) Collective action initiatives to improve marketing performance: lessons from farmer groups in Tanzania. Food Policy 34(1):53–59

Barrett C (2008) Smallholder market participation: concepts and evidence from eastern and southern Africa. Food Policy 34:299–317

Bernard T, Spielman D (2009) Reaching the rural poor through rural producer organizations? A study of agricultural marketing cooperatives in Ethiopia. Food Policy 34(1):60–69

Bernard T, Taffesse A (2012) Measuring aspirations: discussion and example from Ethiopia. International Food Policy Research Institute Discussion Paper 01190

Bernard T, Collion M, de Janvry A, Rondot P, Sadoulet E (2008) Do village organizations make a difference in African rural development? A study for Senegal and Burkina Faso. World Dev 36 (11):2188–2204

Binswanger HP, Nguyen T (2005) Scaling up local and community-driven development: a step-by-step guide. World Bank, Washington, DC

Boyana T, Tshuma M (2013) A review of the possibilities of alleviating poverty and food insecurity challenges in South Africa through agricultural cooperatives. Afr J Agric Res 8 (16):1340–1349

Chirwa E, Matita M, Dorward A (2010) Factors influencing access to agricultural input subsidy coupons in Malawi. Future Agricultures Consortium, Brighton

Coleman J (1966) Foundations for a theory of collective decisions. Am J Sociol 71(6):615–627

de Janvry A, Fafchamps M, Sadoulet E (1991) Peasant household behavior with missing markets: some paradoxes explained. Econ J 101(409):1400–1417

Debrah S, Nederlof S (2002) Empowering farmers for effective participation in decision-making. International Fertilizer Development Center (IFDC) publication IFDC-G-1

Deng W, Hendrikse G (2013) Social capital and incentives in the provision of product quality by cooperatives. In: 140th EAAE seminar: Theories and empirical applications on policy and governance of agri-food value chains, Perugia, Italy

Dorward A, Kydd J (2004) The Malawi 2002 food crisis: the rural development challenge. J Mod Afr Stud 42(3):343–361

Fafchamps M, Hill R (2005) Selling at the farm-gate or travelling to market. Am J Agric Econ 87 (3):717–734. Agricultural and Applied Economics Association

Fafchamps M, La Ferrara E (2012) Self-help groups and mutual assistance: evidence from Urban Kenya. Econ Dev Cult Chang 60(4):707–733. University of Chicago Press

Faysse N, Srairi M, Errahj M (2012) Local farmers' organisations: a space for peer-to-peer learning? The case of milk collection cooperatives in Morocco. J Agric Educ Ext 18 (3):285–299

Fischer E, Qaim M (2011) Smallholder farmers and collective action: what determines the intensity of participation? In: Proceedings of the German development economics conference, Berlin 2011, No. 28

Fischer E, Qaim M (2012) Linking smallholders to markets: determinants and impacts of farmer collective action in Kenya. World Dev 40(6):1255–1268

Goetz S (1992) A selectivity model of household food marketing behavior in Sub-Saharan Africa. Am J Agric Econ 74(2):444–452

Hernández-Espallardo M, Arcas-Lario N, Marcos-Matás G (2013) Farmers' satisfaction and intention to continue membership in agricultural marketing cooperatives: neoclassical versus transaction cost considerations. Eur Rev Agric Econ 40(2):239–260

Holmen H (1990) State, cooperatives and development in Africa, Scandinavian Institute of African studies. University of Uppsala, Uppsala

Hussain M (2014) The role of cooperative organizations in rural community development in Nigeria: prospects and challenges. Acad Res Int 5(3):3–21

Hussi P, Murphy J, Lindberg O, Brenneman L (1993) The development of cooperatives and other rural organizations: the role of the World Bank. World Bank, Washington, DC

Krishna A (2002) Active social capital: tracing the roots of development and democracy. Columbia University Press, New York

Levi Y, Davis P (2008) Cooperatives as the 'enfants terribles' of economics: some implications for the social economy. J Socio-Econ 37(6):2178–2188

Macharia M, Mshenga P, Ngigi M, Gido O, Kiprop K (2014) Effect of transaction costs on smallholder maize market participation: case of Kwanza District, Trans Nzoia County, Kenya. Int J Dev Sustain 3(4):715–725

Makundi E, Mathenge M, Ngigi M (2013) Sweet potato marketing among smallholder farmers: the role of collective action. In: 4th International Conference of the African Association of Agricultural Economists (ICAAAE), September 2013, Hammamet, Tunisia

Mujawamariya G, D'Haese M, Speelman S (2013) Exploring double side-selling in cooperatives: case study of four coffee cooperatives in Rwanda. Food Policy 39(4):72–83

Mulangu F (2011) Selling at the farm-gate or traveling to the market: a conditional farm level model. J Dev Areas 44(2):95–107

Muradian R (2013) The potential and limits of farmers' marketing groups as catalysts of rural development. In: UNRISD conference: Potential and limits of social and solidarity economy, 6–8 May, 2013, Geneva

Nilsson J, Svendsen GLH, Svendsen GT (2012) Are large and complex agricultural cooperatives losing their social capital? Agribusiness 28:187–204

Ntale FK (2013) Forcing the dumb to sing? Rejuvenating farmer cooperatives in Uganda amidst waning social capital. J Afr Stud Dev 5(8):224–235

Obademi O, Ogbuji I (2013) Leveraging on rural cooperatives in informal sector financing for economic growth. Res J Finance Account 4(5). Department of Finance, University of Lagos, Nigeria

Olson M (1965) The logic of collective action: public goods and the theory of groups. Harvard University Press, Cambridge, MA

Ostrom E (1994) Constituting social capital and collective action. In: Workshop on political theory and policy analysis: heterogeneity and collective action, W 93-18. Indiana University, Bloomington, Indiana

Poole N, de Frece A (2010) A review of existing organisational forms of smallholder farmers' associations and their contractual relationships with other market participants in the East and Southern African ACP region. Food and Agriculture Organization of the United Nations (FAO), AAACP Paper Series, no. 11, January 2010

Pretty J, Ward H (2001) Social capital and the environment. World Dev 29(2):209–227

Reardon T, Berdegué J (2002) The rapid rise of supermarkets in Latin America: challenges and opportunities for development. Dev Policy Rev 20(4):371–388

Reardon T, Barrett C, Berdegué J, Swinnen J (2009) Agrifood industry transformation and small farmers in developing countries. World Dev 37(11):1717–1727

Reuben E (2003) The evolution of theories of collective action. Master's thesis, Tinbergen Institute

Rogasa C, Golan J (2012) The role of rural producer organizations for agricultural service provision in fragile states. International Food Policy Research Institute (IFPRI), Washington, DC

Ruben R, Heras J (2012) Social capital, governance and performance of Ethiopian coffee cooperatives. Ann Public Coop Econ 83(4):463–484

Salifu A, Lee R, Keefe M, Kolavalli S (2010) Farmer based organizations in Ghana. Ghana Strategy Support Program, IFPRI

Shiferaw B, Obare G, Muricho G, Silim S (2009) Leveraging institutions for collective action to improve markets for smallholder producers in less-favored areas. Afr J Agric Resource Econ (AFJARE) 3(1):1–18

Staal S, Chege L, Kenyanjui M, Kimari A, Lukuyu B, Njubi D, Owango M, Tanner J, Thorpe W, Wambugu M (1998) Characterisation of dairy systems supplying the Nairobi milk market: a pilot survey in Kiambu District for the identification of target groups of producers. Smallholder Dairy (R&D) Project, p. 85. KARI/MoA/ILRI Collaborative Dairy Research Program, International Livestock Research Institute (ILRI), Nairobi, Kenya. http://www.cgiar.org/ilri/info/databases.cfm

Uphoff N, Wijayaratna C (2000) Demonstrated benefits from social capital: the productivity of farmer organizations in Gal Oya, Sri Lanka. World Dev 28(11):1875–1890

Wanyama FO, Develtere P, Pollet I (2009) Reinventing the wheel? African cooperatives in a liberalized economic environment. CoopAFRICA working paper 1, Dar es Salaam, Tanzania. International Labor Organization (ILO)

Verhofstadt E, Maertens M (2013) Cooperative membership and agricultural performance: evidence from Rwanda. Bioeconomics Working Paper Series 6, Division of Bioeconomics, Department of Earth and Environmental Sciences, University of Leuven, Belgium

Chapter 10
To Be or Not to Be a Member of a Grassroots Institution: Evidence from a Social Network Analysis in Rural Areas of Ghana

10.1 Introduction

Since the seminal work of Olson (1965), social interactions have become an increasingly influential component of economic reasoning. While not much was written on the topic in the immediate aftermath of publication, Olson's basic reasoning lies at the root of the recent rapid expansion of social-network literature in economics. Brock and Durlauf (2001) and Manski (1993, 2000) survey a range of contexts in which social networks have been used to explain individual and aggregate outcomes in economics. In the case of developing countries, Mulangu (2014) points out that social networks are important catalysts—influencing the efficiency of exchanges; serving as insurance, especially in the realm of microfinance; and facilitating marketing and technology transfer in agrarian economies.

Social networks also affect meso-level entities, such as enterprises, and their performance. Evidence from Ghana shows that large entrepreneurs maintain large innovation networks, suited to providing information about technology and markets, while smaller entrepreneurs keep small networks suited to reducing information asymmetries and thus supporting informal credit and risk-sharing arrangements (Barr 2002). In addition, evidence from Burkina Faso reveals that informal entrepreneurs must combine strong and wide social support ties with weaker business ties in order to be successful (Berrou and Combarnous 2012).

Regarding the labor market, further evidence from Burkina Faso stresses the importance of kinship in starting a business. In the same vein, a larger network increases the probability of finding wage employment rather than self-employment, thanks to better access to information. Weakness of ties between siblings also exerts

This chapter was written by Vera Chiodi (Sorbonne University Paris 3) and Francis Mulangu (ACET).

N. Depetris Chauvin et al., *Agricultural Supply Chains, Growth and Poverty in Sub-Saharan Africa*, Advances in African Economic, Social and Political Development, DOI 10.1007/978-3-662-53858-6_10

a positive effect on job-seeking, as it increases the likelihood that the seeker will maintain more non-family contacts and so broaden his or her business network. Stronger sibling ties do not necessarily help the unemployed find jobs, as they may exert a disincentive effect through the provision of a familial safety net, or a pressure to redistribute earnings among the family, thus limiting job-seeking efforts (Pasquier-Doumer and Nordman 2013).

Recent years have witnessed a renewed policy interest in community-based development, predicated on the premise that interventions at the level of a local community can deliver more effective and more equitable development (World Bank 2002). In practice, interventions are often channeled through grassroots institutions (GRIs), especially in rural areas. Seeking to describe the composition and characteristics of GRIs, Arcand and Fafchamps (2012) found robust evidence of ethnic proximity and similar wealth and household size among members of GRIs. This assortative matching, they claim, may undermine GRIs' effectiveness in improving inclusive welfare if these characteristics differ from those prevalent in the community where members live.

Fafchamps and La Ferrara (2012) studied the role of urban GRIs in helping the poor manage risk by examining the incomes of individuals who have joined informal GRIs in various poor neighborhoods of Nairobi. The goal of this research was to understand the extent to which GRIs help to pool income and hence serve as a form of insurance in an urban context. Results show that the incomes of individuals in the same group are more correlated among each other than the incomes of individuals in different groups, and that this is not driven by assortative matching into groups. These findings are consistent with the idea that GRIs play not only an insurance role but a redistributive one as well.

Among the growing literature on community-driven interventions that promote the use of GRIs to deliver development programs, Desai and Joshi (2014) evaluate the impact of an intervention that organized women farmers into rural producer GRIs in India. The initiative provided training, information, access to inputs, risk mitigation, and market linkages over a period of 18 months. Through a randomized evaluation, the authors found that the program exerted a positive impact, particularly for landless women.

Many studies have contributed to our understanding of the role of institutions in poor settings, as illustrated above in the case of GRIs. However, there is still scant empirical evidence on the way local and relatively small institutions like GRIs foster the welfare of their members through the acquisition of social networks. The present work contributes to this emerging body of literature, and, to our knowledge, is the first one to provide empirical evidence.

In a departure from previous approaches, we study social networking inside and outside a community within the framework of an established local institution, the GRI. The underlying argument is that within this formal institution, GRI member interactions will differ from those of their non-member neighbors, and the difference made by the GRI can help inspire policy recommendations. GRI members' linkages may be enhanced by economies of scale, which will facilitate interactions with outside institutions. In this study, we also indirectly identify the effect of reducing the transaction cost of social networking through GRI membership.

In order to introduce a social network analysis into the role played by GRIs in rural areas of Ghana, this work aims to study the ways in which GRIs interact with links inside and outside the rural community. Quite often, the GRIs emerged at the local (community) level but had links outside the community as well—with other GRIs; with governmental institutions (GIs); with other households; or, if applicable, with their mother GRIs. Consequently, GRI members are expected to have better access to external welfare systems delivered exclusively by a GI or an NGO.

The existence of links between GRIs and households in each village regarding exchanges of credit, inputs, labor, and land is analyzed in the same vein as Jaimovich (2013). Networks inside the village can be interpreted as a community-driven development process, or as a spillover effect when the internal link involves both a GRI member and a non-member. Networks outside the village, on the other hand, can be interpreted as external market access. In particular, we test the degree of complementarity or substitutability between internal and external links in order to inform our ideas of how households and GRIs interact. The main hypothesis of substitutability between internal and external links is tested under the assumption that households with external links are less likely to be involved in economic transactions within the village. For each network, the potential marketability through external links will tend to be relatively inelastic (or elastic) if the supply capacity is constrained at the local level (or unconstrained).

Using a mix of qualitative and quantitative information, the empirical analysis provides information on an additional level: the correlation between economic indicators and membership in the GRI versus non-membership, taking into account links with potential governmental institutions belonging to the same networks. For instance, the secondary hypothesis will test the extent to which GRIs play a role vis-à-vis governmental institutions in the supply chain of crop production, storage, and marketability, controlling for the GRIs' external links (i.e., having a mother GRI). The underlying idea is that GRIs may emerge with or without governmental institutions.

Following this introduction, this chapter is divided into four sections: Sect. 10.2 presents the data collection and sampling strategy and the descriptive statistics of our sample; Sect. 10.3 explains the methodology used; Sect. 10.4 discusses the econometric results; and Sect. 10.5 concludes and suggests further research.

10.2 The Data

We present a new data set of social and economic networks, collected in nine rural communities in the Tolon-Kumbungu District of northern Ghana between August and September 2013. The district was chosen based on a GRI strength index,[1]

[1]Available from the authors upon request.

created using a database of information on: (1) whether the GRI has a bank account; (2) if the GRI is registered with the Department of Cooperatives, Registrar General's Department, or the Ministry of Food and Agriculture (MoFA); (3) level of the GRI's production (MT), acreage (ha), and yields; (4) the source of its existence; and (5) its main activity (crop production, processing, marketing, livestock production, aquaculture). Based on a list of GRIs registered with MoFA, we have randomly selected 10 GRIs from nine communities. A household survey was conducted using 15 % of the total households in these communities. Additionally, a survey of the secretaries or chairmen of the GRIs was carried out in order to capture potential institutional interactions. In total, 150 households were interviewed, of which 75 are members of a GRI and 75 are not. The non-GRI members interviewed were all neighbors of GRI members, arbitrarily selected by identifying the fifth closest neighbor to each GRI household.

The Tolon-Kumbungu District covers a land area of 2741 km^2, about 3.9 % of the total land mass of the Northern Region of Ghana. The district is bounded by the West Mamprusi District to the north, Central Gonja District to the south, and West Gonja to the west. Tamale, the capital of the Northern Region, and the Savelugu-Nanton District share eastern boundaries with Tolon-Kumbungu. The position of Tolon-Kumbungu relative to the rest of the country is illustrated in Fig. 10.1 below.

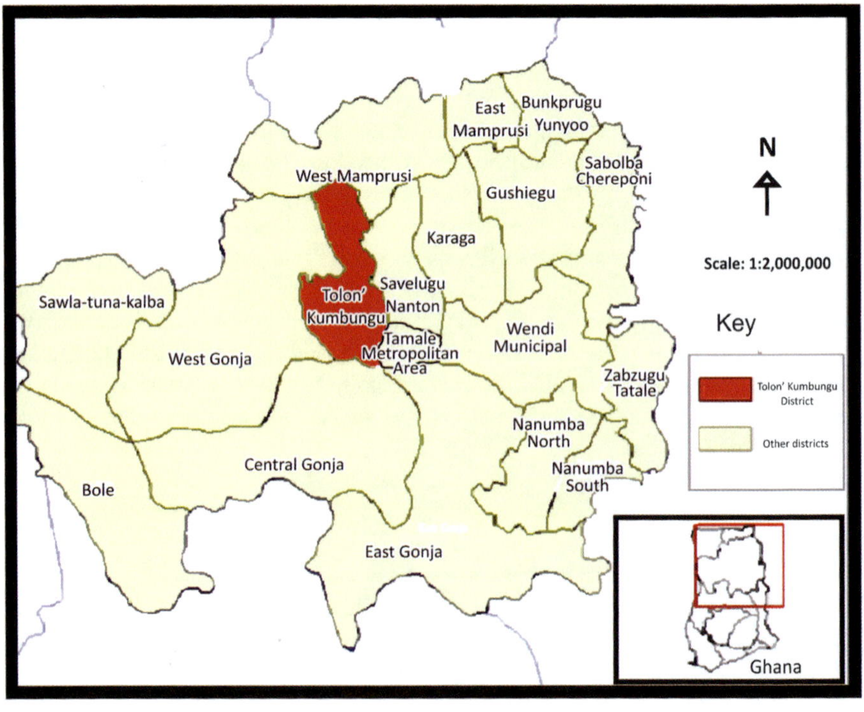

Fig. 10.1 Tolon-Kumbungu District

The district lies in the Guinea Savannah zone, with a soil dominated by clay and loam. Major tree species include nim, baobab, shea, and *dawadawa*. The zone is characterized by a single maximum rainfall season, which starts in April or May, becomes irregular in August, and ceases in October, ushering in the dry season. Mean monthly rainfall ranges from 500 to 1100 mm, and mean monthly temperatures fall between 17 and 40 °C. The population of the district was measured at 132,883 in Ghana's 2000 Population and Housing Census, with a growth rate of 2.7 %. Population density is estimated at 50 inhabitants per square kilometer.

The main agricultural activity in the district is crop farming, with cereal crops dominating. The main crops grown are maize, rice, millet, yams, cowpeas, pigeon peas, and soybeans. Vegetable crops grown include tomatoes, peppers, okra, onions, and eggplant, whilst cash crops include cotton and groundnuts. Other agricultural activities include livestock production, especially breeding guinea fowl. In our sample, we found that the dominant crop produced by respondents was groundnuts, followed by maize and rice. More than 68 % of the respondents produce primarily groundnuts, and the difference between GRI members and non-members is negligible.

Although the Tolon-Kumbungu District is mainly agrarian, with 81 % of residents' income coming from agriculture, the population is also engaged in informal, non-agricultural economic activities. These include hospitality services in restaurants and guesthouses, petty trading, and shea butter and groundnut oil extraction, all dominated by women; and blacksmithing, bicycle repair, and fitting services, which are dominated by men.

Using records from registered farmer-based organizations at MoFA, we found that 29 GRIs are registered in the Tolon-Kumbungu District. The average number of members per GRI is about 25, with an average of 14.3 males and 10.7 females. Market challenges GRIs seek to address can be grouped into the crop production, processing, and/or marketing categories. Based on these records, we estimated that 83 % of the GRIs in the district seek to address production-related market failures, while 21 % and 3 %, respectively, seek to address market failures associated with processing and marketing.

10.3 Sampling Procedure

The data used for the empirical analysis was collected after a two-stage sampling procedure. First, using national census data on all GRIs registered in the Northern Region and their attributes, we estimated their strength based on a number of variables, including ownership of a bank account, number of places where the GRI was registered, and evidence of meeting minutes. After identifying Tolon-Kumbungu's GRIs from this first step, we randomly selected 10 of the 29 registered GRIs, based in nine communities, as listed below in Table 10.1.

The average number of households in each community is 139; the largest one, Tolon, has more than 400 households, and the smallest, Bogunaayili, has

Table 10.1 Registered GRIs in 10 selected communities and total number of households

Community (village) name	GRI name	Number of members in GRI	Total number HH in community
Tali Botingli	Wumpini Farmers Group	25	32
Yobzeiri	Tiyumtaba	49	45
Tibogu-Naayili	Suhuyini Women Group	58	112
Tolon Gbambaya	Bobgu Viela Farmers	40	40
Woriobogu Kamonaayili	Suglo Mbori Buni	41	60
Bogunaayili	Bang Biebu Farmers	42	20
Tibung	Tiyumtaba	49	225
Kumbung Kukuo	Suglo Mbori Buni	50	54
Tolon	Suglo Mbori Buni	55	400
	Bukuru Sung Women Groups	30	
TOTAL		439	988[a]

[a]Please note here that two GRIs (Suglo Mbori Buni and Bukuru Sung Women Groups) were taken from similar community (Tolon)

20 households. Before initiating the data collection, we visited the communities where the GRIs are based to introduce our project to the GRI leaders and request their lists of members, which we used to randomly select GRI members for interviewing. The sample size for each GRI, however, was chosen in proportion to the size of the communities in which the GRIs are based so that at least 5 % of the community was captured. To identify the corresponding non-GRI members, we chose a household located five houses from each selected GRI member's house in the community, a second-best approach given that most communities did not have census figures on their households.

In the end, we interviewed about 150 households, corresponding to more than 15 % of the total population in all the communities we visited. Table 10.2 presents the number of GRI members and non-GRI members chosen in each community. The survey we gave consists of two sections: first, a standard household questionnaire designed to collect a vector of household/GRI characteristics on agricultural activities; second, a set of questions designed to understand the economic and social networks. The networks data will be used later to construct networking matrices for households within and between communities (including both GRI members and non-members) in four categories: credit, input, labor, and land. An illustration of the relative position of each community is presented in Fig. 10.2.

Table 10.2 Total numbers of non-GRI respondents and GRI respondents

Community (village) name	GRI name	Total number of respondents	Number of GRI member respondents
Tali Botingli	Wumpini Farmers Group	6	3
Yobzeiri	Tiyumtaba	5	3
Tibogu-Naayili	Suhuyini Women Group	18	9
Tolon Gbambaya	Bobgu Viela Farmers	8	4
Woriobogu Kamonaayili	Suglo Mbori Buni	8	4
Bogunaayili	Bang Biebu Farmers	9	5
Tibung	Tiyumtaba	35	17
Kumbung Kukuo	Suglo Mbori Buni	5	3
Tolon	Suglo Mbori Buni	31	15
Tolon	Bukuru Sung Women Groups	25	12
TOTAL		150	75

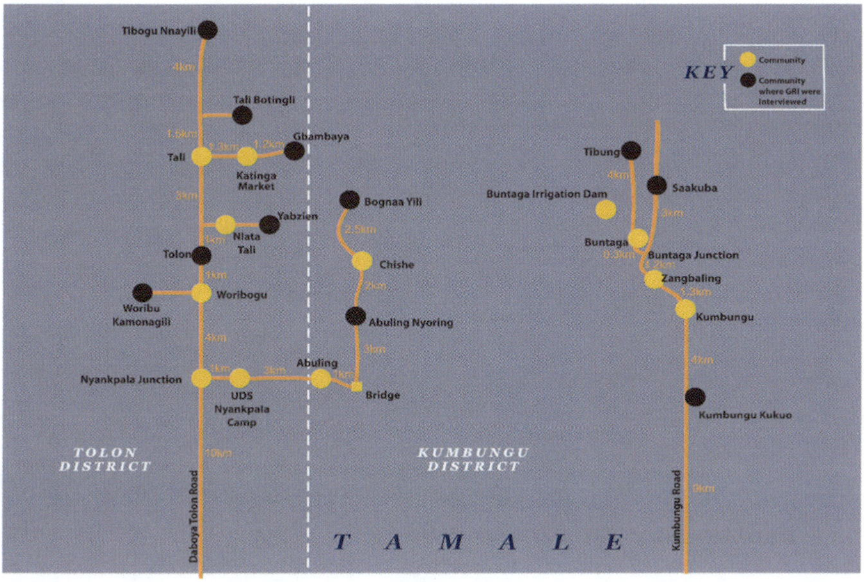

Fig. 10.2 Location of the GRIs. Source: Authors'construct

10.4 Descriptive Statistics

According to the qualitative data, the main objectives of the creation of the GRIs can be summarized as follows: to generate mutual assistance on the farm; to secure support (credit, training, subsidies) from NGOs and government; to pool resources for crop and cattle production; to develop agribusiness; and to access (formal) credit and technical knowledge from ministries or other governmental institutions (GIs).[2] Table 10.3 reports summary statistics describing the 10 GRIs in our sample, which differ in size and longevity. The smallest group had 25 members and the largest had 58, with an average of 43.9 members per group. On average, the groups have been in place for 5.1 years; the newest is 3 years old, and the oldest is 12. Recruitment into the groups is not always based on skill or productivity, but on a person's willingness to obey the rules and general disposition toward trustworthiness and hard work, qualities to be informally evaluated by current group members. GRIs seem to actively engage with both other GRIs and GIs outside of the community. On average, a GRI interacts three times a year with another GRI and about four times with GIs outside its community. Most the GRIs require periodic fees from their members, in considerably variable amounts, from US$1 to US$25 per year. Four out of the ten surveyed GRIs are branches of larger GRIs based outside of the community. The main crop marketed is maize.

Turning to individual characteristics (Table 10.4), we found that the average household size is about 12 individuals, and the average age of the household head is 49 years old. Only 18 % of the sample households are headed by a female, and 80 % of the respondents are illiterate. The average land size of each household is 13 acres, and most farmers believe that the quality of their land is low. Average agricultural income stands at about US$793 per annum, and average non-agricultural income is US$187 per annum, or 19 % of total income.

Table 10.3 Descriptive statistics of GRIs

Variable	Mean
Group size	43.9
Years in operation	5.1
Number of yearly meetings with other GRIs	3
Number of yearly meetings with GIs	4
Number of GRI members presently	44.7
Number of female members	24.6
Number that are branches of larger GRIs	4
Total number of households in community	141.8
Number of GRIs contacted by GRI in the last year	2.8
Number networked with other community GRIs	0.3
What crop is at the center of your organization?	Maize

[2]This data is taken from the qualitative surveys addressed to the executives of the 10 GRIs.

Table 10.4 Household characteristics

Variable	Mean/%
Household size	12
Household head age	49
Household headed by female	18 %
Land size (in acres)	13
Illiteracy rate	80 %
Farmer with low quality land	80 %
Non-farm income (in USD)	187
Farm income (in USD)	793

Table 10.5 Nature and strength of social networking

Variable	Mean	Standard deviation
Land traded **outside** community (acres)	0.49	1.66
Land traded **inside** community (acres)	1.42	2.41
Labor traded **outside** community (days)	7.0	45.06
Labor traded **inside** community (days)	12.10	37.06
Number of inputs traded **outside** community	0.46	0.73
Number of inputs traded **inside** community	0.5	0.92
Credit traded **outside** community (in USD)	14.20	58.28
Credit traded **inside** community (in USD)	38.60	87.58

Farmers traded credits, inputs, labor, and land with other farmers and institutions inside and outside their communities. We estimated that farmers (both GRI members and non-members) traded on average about 0.5 and 1.4 acres of land with external and internal entities, respectively. In the case of labor, we found that farmers traded 12 days of work within their communities, compared with 7 days of work outside the community. Although we have not examined it explicitly, one of the reasons behind this could be the cost of traveling outside the community, thus restricting the reach of a farmer's labor trade. Further, Table 10.5 (below) shows that farmers traded on average 0.5 inputs inside their communities, compared with 0.46 inputs outside. The credit figures show a far more significant gap: Farmers traded US$38.60 inside their communities versus US$14.20 outside.

The description provided in Table 10.5 suggests that farmers interact more inside the community than outside of it for the four links considered here. This is anticipated, since interacting outside the community is associated with a higher transaction cost.

10.5 Empirical Strategy

We estimated a set of variables measuring the strength of the network links for the different respondents relative to credit, land, input, and labor linkages. We then divided the network variables into links outside and inside of the community. For credit, land, and input networks, we created additional variables to capture whether the trading transaction consisted of borrowing or lending the resources. In the end, we constructed and evaluated more than 30 network variables and compared the differences between GRI members and non-GRI members.

The identification strategy is based on a comparison of means of outcome variables, Y, by GRI group membership, G. We estimate the following equation for each household:

$Y_i = \alpha + \beta G_i + \gamma Var_i + \varepsilon_i$ with $E\ (\varepsilon_i/G_i) = 0$, meaning that the error term is uncorrelated with G_i. The proposed analysis implies that:

$E\ (Y_i/G_i = 1)$ is the average outcome in the GRI member group.

$E\ (Y_i/G_i = 0)$ is the average outcome in the non-GRI member group.

We then estimate the difference: $\Delta^{Diff} = E\ (Y_i/G_i = 1) - E\ (Y_i/G_i = 0)$

Our basic assumption, then, is that:

$E\ (Y_i(0)/G = 1) = E(Y_i(0)/G = 0)$

where Y is the outcome variable and G is the independent variable of being ($G = 1$) or not being ($G = 0$) a GRI member. We test this assumption over three categories of variables:

1. Total, internal, and external links for land, credit, input, and labor.
2. Economic and perceived well-being indicators.
3. Potential constraints faced by farmers in three stages of the value chain of the crops, mainly production, storage, and marketing.

The coefficient is the difference of means, as depicted in Fig. 10.3, and we report the t-test of difference of means. Three models are tested: (a) Model 1, without controls; (b) Model 2, with control variables *Var* (marriage and kinship links, as well as a dummy indicating whether the GRI has a mother GRI; and (c) Model 3, which adds demographic control variables such as household size and household head education.

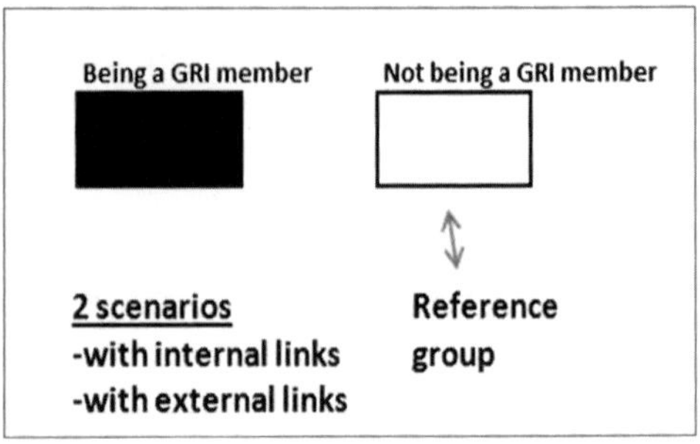

Fig. 10.3 The methodology: comparison of means of two groups. Source: Authors

10.6 Interpretation of Results

After an average of 4 years of active membership, our results show no systematic differences between GRI members and non-GRI members. However, some relevant variables within our three categories depict a few interesting results, discussed below.

10.6.1 Total, Internal, and External Links for Land, Credit, Input, and Labor

Our main result is that GRI membership helps significantly to relax credit and input constraints, as shown in Table 10.6. Both corresponding coefficients are significant, at the 5 or 10 % level across all specifications. This suggests that there is a significant advantage for GRI members in access to the input network (borrow >

Table 10.6 Network regression results for four categories

Variable	GRI	Non-GRI	Diff	Effect M1	Diff	Effect M2	Diff	Effect M3
Has a land network	0.55	0.52	0.03		0.03		0.02	
Has a labor network	0.63	0.51	0.12		0.12		0.11	
Has an input network	0.71	0.55	0.16	**	0.16	*	0.15	*
Has a credit network	0.6	0.44	0.16	*	0.17	**	0.16	*
Has an internal land network	0.47	0.35	0.12		0.12		0.12	
Has an internal labor network	0.52	0.4	0.12		0.12		0.11	
Has an internal input network	0.32	0.24	0.08		0.09		0.09	
Has an internal credit network	0.4	0.4	0		−0.01		−0.01	
Has an external land network	0.08	0.17	−0.09	*	−0.09		−0.1	*
Has an external labor network	0.11	0.11	0		0		0	
Has an external input network	0.39	0.31	0.08		0.06		0.07	
Has an external credit network	0.2	0.04	0.16	***	0.17	***	0.17	***

Note: The second and third columns give the mean value of the outcome variable in the GRI member group and in the non-GRI member group, respectively. The Diff columns represent the difference between groups across the three models: M1, M2, and M3. Differences are significant at 1 % (***), 5 % (**), and 10 % (*). See text for the definition of each model

lend)[3] and the credit network (borrow > lend),[4] both internal and external, over non-GRI members. Membership increases by around 16 % points the probability of access to both credit and inputs (such as fertilizer, tools, or equipment). In the case of credit, this result is driven mainly by the credit external link (borrow > lend),[5] particularly as it is significant at the 1 % level for all three models.

Today in Ghana, as in many other developing countries, GRIs are increasingly replacing individual farmers as the preferred clients of microfinance institutions (MFIs).[6] Group lending, executed via GRIs, has become the norm among rural lenders as a means of reducing loan default through peer pressure. For that reason, if a farmer seeks access to a credit facility, s/he is better off joining a GRI; this is especially true if the farmer is poor and has no assets to use as collateral, as group solidarity is a form of insurance to the lender. Although we will find in the next section that GRI members tend to have lower income than non-GRI members, the higher total credit links enjoyed by members, as observed above, are driven by external credit links made possible by GRIs.

Since 2008, the government of Ghana has used private input dealers to distribute input subsidies to farmers in response to the food crisis. This was meant to be a one-time policy, but quickly became permanent, and its implementation strategy has since evolved. Fertilizer subsidies from the government of Ghana covered 36 % of the retail price. To reduce the transaction cost for delivery of inputs, and to help poor farmers who may not be able to afford the remaining 64 % of the retail fertilizer price by themselves, farmers were encouraged to form groups to help address these constraints and ensure their access to the inputs. Our results support this, and it is not a new practice in Ghana. Historically, many post-independence African governments have used GRIs as preferential channels for the provision of credit, often linked to the distribution of agricultural inputs (Hussi et al. 1993; Debrah and Nederlof 2002). Farmers were often coerced to join GRIs, and membership in GRIs was usually a prerequisite for obtaining credit from the government (Holmen 1990). However, realizing that farmers saw GRIs mainly as a means to secure public support, and not to promote competitive agribusiness, the government stopped supporting GRIs in the 1980s. It is only in recent years that channeling development support via GRIs has attracted new interest (Binswanger and Nguyen 2007), predicated on the premise that interventions at the local community level can deliver more effective and equitable development. This view clearly diverges from the earlier top-down approach, which holds a central entity responsible for

[3]Borrowing transactions dominate over lending ones. Results are not shown, but available upon request.

[4]Ibid.

[5]Ibid.

[6]This statement is supported by the fact that more than 80 % of rural agricultural credit in northern Ghana is given to groups as opposed to individual farmers. This conclusion was drawn after surveying all rural banks operating in northern Ghana.

implementing a development project and has been criticized as elitist (Binswanger and Nguyen 2007).

Land and labor links, on the other hand, did not appear to differ significantly between GRI and non-GRI members. However, and in contrast to our previous result, non-GRI members seemed to then compensate for their lack of credit and input resources with marginally higher (as the coefficient is only significant at the 10 % level) external land networks, meaning they have a land lending or borrowing relationship with someone outside the community. The difference is about 9 % points. It is important to remember here that having an external land network does not imply land tenure; for instance, non-GRI members surveyed did not have significantly more land than GRI members, as shown in Table 10.7. The land market in Ghana is weak: rural lands are controlled by village chiefs or traditional leaders, who decide who gets a land lease and for how long. For that reason, belonging to a GRI does not improve one's access to land.

In the case of labor links, farmers may get labor via their household members and children, so the labor gain through GRIs is not that attractive, especially considering that the household size in our sample averages about 12. In addition, households in our surveyed area share compounds with other households, a system established in order to facilitate mutual assistance and sharing of other services provided within the community. Thus, labor links are similar between GRI and non-GRI members because labor is readily accessible.

In parallel, it does not seem that GRI members interact differently (or even more or less) with the rest of the community or with outside village households than do non-GRI members.

10.7 Indicators of Economic and Perceived Well-Being

The significant difference in total household income between the two groups indicates that GRI members tend to be poorer than non-GRI members. This seems to counter Bernard and Spielman (2009), Francesconi and Heerink (2010), and Fischer and Qaim (2012), whose results imply that GRIs tend not to serve the smallest growers that make up the poorest of the poor. This discrepancy could be because the GRIs surveyed here were largely established in an area that facilitates development program delivery to poor farmers; meanwhile, those studied in the cited literature were set up to address marketing-related challenges, a problem mostly faced by active and dynamic farmers, who tend to be richer.

As explained earlier, our sample is composed of households that actively participate in agricultural activities. We might have anticipated that agricultural income for GRI members would be higher than that of non-members; however, our results do not validate this assumption. Given that the means of agricultural income do not differ significantly between the two groups, as shown in Table 10.7, our survey outcome suggests that most GRIs in Ghana fail to address marketing problems faced by their members. This conclusion, also illustrated by Salifu et al.

Table 10.7 Variable regression results of economic and own perceived well-being

Variable	GRI	NGR	Diff	Effect M1	Diff	Effect M2	Diff	Effect M3
Land owned (acres)	13.65	13.68	−0.03		−0.03		−0.64	
Non-ag income	264.36	599.29	−334.93	**	−325.29	**	−327.04	**
Agri income 2011–2012	1641.96	2006.97	−365.01		−342.11		−384.93	
Sold cash crop to trader	0.81	0.81	0.00		0.01		0.01	
HH member migrated	0.59	0.64	−0.05		−0.01		−0.02	
Consider relatively wealthy	0.37	0.21	0.16	**	0.14	*	0.13	*
Satisfied with agri production	0.27	0.40	−0.13	*	−0.10		−0.09	
Satisfied with post-harvest	0.35	0.44	−0.09		−0.06		−0.06	
Satisfied with input costs	0.11	0.07	0.04		0.05		0.05	
Satisfied with market prices	0.20	0.15	0.05		0.09		0.09	
Better position production	0.25	0.17	0.08		0.06		0.05	
Better position post-harvest	0.16	0.19	−0.03		−0.04		−0.04	
Better position costs	0.04	0.11	−0.07		−0.06		−0.06	
Better position prices	0.05	0.12	−0.07		−0.08	*	−0.08	*
% prod sold at the market	56.80	59.93	−3.13		−1.93		−1.98	

(continued)

Table 10.7 (continued)

Variable	GRI	NGR	Diff	Effect M1	Diff	Effect M2	Diff	Effect M3
% prod sold in village	16.63	15.48	1.15		1.46		1.31	
% prod consumed	22.24	24.08	−1.84		−3.39		−3.22	

Note: The second and third columns give the mean value of the outcome variable in the GRI member group and in the non-GRI member group, respectively. The Diff columns represent the difference between groups across the three models: M1, M2, and M3. Differences are significant at 1 % (***), 5 % (**), and 10 % (*). See text for the definition of each model

(2010), would explain the strong desire of GRI members to fetch better prices for their products. Bernard et al. (2008a) and Francesconi and Heerink (2010) also show that overall, cooperatives in Ethiopia can offer better prices, but have a limited capacity to raise levels of market integration or commercialization, particularly among the smallest farmers. All this demonstrates that although access to credit and inputs are the objectives of GRI, they may not necessarily translate in practice into agricultural revenue.

On the other hand, GRI membership provides farmers with a sense of confidence and awareness. GRI member households described themselves as wealthier than other households in their communities, but less satisfied with their level of production than non-GRI members and with the prices their products fetched, as noted above. GRI members probably consider themselves wealthier than non-GRI members because they can rely on external supports they receive through their organizations. Similarly, this external interaction via GRI makes them more aware of the price distribution outside their communities and in the capital, Tamale, where prices are higher than in the Tolon-Kumbungu District.

10.8 Production, Storage, and Marketing Stages for Diverse Constraint Variables

A few significant differences among the constraint variables were identified between the two groups surveyed (see Table 10.8). GRI members are more prone to identify their own constraints, suggesting that GRI members may acquire more knowledge about prices and market structure than non-GRI members. Again, this is probably because GRI members interact more with external GRIs, GIs, and other NGOs that support them through various development programs.

These results are partially consistent with Bernard et al. (2008a) and Bernard and Spielman (2009), who found that GRIs function as vital conduits for marketable surpluses of agricultural commodities, agricultural inputs, credit technologies, organizational innovations, institutional innovations, and capacity strengthening.

Table 10.8 Variable results of potential constraints related to production, storage, and marketing stages

Variable	GRI	Non-GRI	Diff	Effect M1	Diff	Effect M2	Diff	Effect M3
Production stage								
Input cost constraint to you	1.00	0.99	0.01		0.01		0.01	
Weather uncertainty constraint	0.72	0.60	0.12		0.10		0.10	
Credit constraint to you	0.69	0.57	0.12		0.13		0.12	
Price uncertainty constraint	0.63	0.55	0.08		0.07		0.06	
Degraded soil constraint to you	0.73	0.71	0.03		0.03		0.03	
Storage stage								
Inadequate infrastructure constraint	0.67	0.61	0.05		0.05		0.04	
High storage cost constraint	0.25	0.16	0.09		0.06		0.06	
Price volatility constraint	0.29	0.13	0.16	**	0.15	**	0.14	**
Risk of theft constraint to you	0.23	0.08	0.15	**	0.15	**	0.14	**
Pressing need of cash constraint	0.63	0.49	0.13		0.15	*	0.15	*
Transaction stage								
Transaction cost constraint	0.27	0.21	0.05		0.02		0.02	
Low bargaining power constraint	0.67	0.59	0.08		0.06		0.06	
Information asymmetry constr.	0.39	0.23	0.16	**	0.12		0.11	
Low farm gate price constraint	0.09	0.03	0.07	*	0.07	*	0.07	*
Low quality of product constr.	0.07	0.05	0.01		0.02		0.01	

Note: The second and third columns give the mean value of the outcome variable in the GRI member group and in the non-GRI member group, respectively. The Diff columns represent the difference between groups across the three models: M1, M2, and M3. Differences are significant at 1 % (***), 5 % (**), and 10 % (*). See text for the definition of each model

Bernard et al. (2008b) also point out that market-oriented farmers' organizations in Senegal and Burkina Faso are relatively good at providing information and advice to their members. In its Growth and Poverty Reduction Strategy (2006–2009) and Food and Agriculture Sector Development Policy (2007), the government of Ghana emphasized strengthening GRIs' capacity and organizing farmers' organizations and cooperatives to improve coordination among rural smallholders (Republic of Ghana 2005, 2007). Our results, however, suggest the need for additional programs or improvements in existing programs, which could be facilitated by government institutions to help GRIs overcome the cited constraints regarding prices and market structure. Government institutions should maintain stronger connections with GRIs and thus play a more important role in their activities—for example, by providing some type of insurance to help farmers cope with volatile prices, or by building a more transparent market structure.

10.9 Conclusion

This chapter aims at describing the potential extended links of GRI members versus non-GRI members and the resulting economic indicators. Taken together, our results show no systematic differences between the group of households that have a GRI membership and the households without a GRI membership. However, GRI members have acquired significantly more important credit and input connections than their non-GRI neighbors, resulting in greater confidence in their own wealth and increased awareness of prices and market conditions.

We conclude that GRIs help farmers meet their needs primarily by improving their access to credit and inputs (such as fertilizer and tools) that can only be acquired outside their communities, especially if the community in question is poor. On the other hand, GRIs seem to fail in meeting their needs for land and labor. The absence of significant external land and labor links for GRI members may reveal that the supply capacity of land and labor exchanges is highly constrained, suggesting the possibility of further developing fundamental policies in these markets.

The main limitations of our work are two: the small sample presented here (only 15 % of the total population) and the cross-section nature of the data. One avenue for future research could be to increase the number of interviews, which would shed more light on the underlying mechanisms of existing credit and input networks, both internal and external, as well as on the more constrained markets of labor and land in this context.

References

Arcand J-L, Fafchamps M (2012) Matching in community-based organizations. J Dev Econ 98:203–219

Barr AM (2002) The functional diversity and spillover effects of social capital. J Afr Econ 11 (1):90–113

Bernard T, Spielman D (2009) Reaching the rural poor through rural producer organizations? A study of agricultural marketing cooperatives in Ethiopia. Food Policy 34(1):60–69

Bernard T, Collion ME, Janvry A, Rondot P, Sadoulet E (2008a) Do village organizations make a difference in African rural development? A study for Senegal and Burkina Faso. World Dev 36:2188–2204

Bernard T, Tafesse A, Gabre-Madhin E (2008b) Impact of cooperatives on smallholders' commercialization behavior: evidence from Ethiopia. Agric Econ 39:147–161

Berrou J-P, Combarnous F (2012) The personal networks of entrepreneurs in an informal African urban economy: does the 'strength of ties' matter? Rev Soc Econ 70(1):1–30

Binswanger HP, Nguyen T-V (2007) Scaling up local and community-driven development: a step-by-step guide. World Bank, Washington, DC

Brock W, Durlauf S (2001) Discrete choice with social interactions. Rev Econ Stud 168 (2):235–260

Debrah SK, Nederlof ES (2002) Empowering farmers for effective participation in decision-making. African Division and Market Program. International Centre for Soil Fertility and Agricultural Development (IFDC), Lomé

Desai R, Joshi S (2014) Can producer associations improve rural livelihoods? Evidence from Farmer Centers in India. J Dev Stud 50(1):64–80

Fafchamps M, La Ferrara E (2012) Self-help groups and mutual assistance: evidence from urban Kenya. Econ Dev Cult Change 60(4):707–733

Fischer E, Qaim M (2012) Linking smallholders to markets: determinants and impacts of farmer collective action in Kenya. World Dev 40:1255–1268

Francesconi GN, Heerink N (2010) Ethiopian agricultural cooperatives in an era of global commodity exchange: does organizational form matter? J Afr Econ 20:153–177

Holmen H (1990) State, cooperatives and development in Africa, Scandinavian Institute of African Studies. University of Uppsala, Uppsala

Hussi P, Murphy J, Lindberg O, Brenneman L (1993) The development of cooperatives and other rural organizations: the role of the World Bank. World Bank

Jaimovich D (2013) Missing links, missing markets: internal exchanges, reciprocity and external connections in the economic networks of Gambian villages. Munich Personal RePEc Archive Paper 44080, University Library of Munich, Germany

Manski C (1993) Identification of endogenous social effects: the reflection problem. Rev Econ Stud 60:531–542

Manski C (2000) Economic analysis of social interactions. J Econ Perspect 14(3):115–136

Mulangu F (2014) The role of Grassroot Institutions (GRI) in agriculture in Africa. African Center for Economic Transformation working paper

Olson M (1965) The logic of collective action. Harvard University Press, Cambridge

Pasquier-Doumer L, Nordman CJ (2013) Transitions in a West African labour market: the role of social networks. In: Paper provided by Paris Dauphine University in its series Economics Papers from Université Paris—Dauphine, number 12204

Republic of Ghana (2005) Growth and poverty reduction strategy (GPRS II) 2006–2009. Government of Ghana, Accra

Republic of Ghana (2007) Food and agriculture sector development policy (FASDEP II). Government of Ghana, Accra

Salifu A, Francesconi GN, Kolavalli S (2010) A review of collective action in rural Ghana. International Food Policy Research Institute (IFPRI) Discussion Paper 00998

World Bank (2002) Community-driven development: from vision to practice; a technical sourcebook. World Bank, Washington, DC